Ecological Crisis and Water Supply

Brill's Series in the History of the Environment

Series Editor

Aleks Pluskowski (*University of Reading*)

VOLUME 6

The titles published in this series are listed at *brill.com/bshe*

Ecological Crisis and Water Supply

The Case of Andalusia in the Spanish Hydrological Context

Edited by

Juan Manuel Matés-Barco and María Vázquez-Fariñas

BRILL

LEIDEN | BOSTON

Universidad de Jaén

European Union
European Regional
Development Fund

This research is part of the results of the European Regional Development Fund (ERDF)-University of Jaen-1381621 project (2021–2022) entitled "La gestión sostenible de los servicios públicos: Agua en Andalucía (1820–2020)" [Sustainable management of public services: Water in Andalusia (1820–2020)], within the ERDF Operational Programme of Andalusia R+D+i (2014–2020), and has been financed by the Regional Government of Andalusia and the European Regional Development Fund.

Cover illustration: Watch tower of the dam Embalse de los Bermejales, Arenas del Rey, Andalusia, Spain. Photographer: Xavier Espinasse (Jebulon). Wikimedia Commons, CC0 1.0 Universal Public Domain Dedication.

The Library of Congress Cataloging-in-Publication Data is available online at https://catalog.loc.gov

Typeface for the Latin, Greek, and Cyrillic scripts: "Brill". See and download: brill.com/brill-typeface.

ISSN 1876-6595
ISBN 978-90-04-54131-3 (hardback)
ISBN 978-90-04-67963-4 (e-book)
DOI 10.1163/9789004679634

This book is printed on acid-free paper and produced in a sustainable manner.

Contents

Preface VII
Acknowledgements XV
List of Figures and Tables XVI
Notes on Contributors XVIII

1 Water Supply Management in Andalusia 1
Juan Manuel Matés-Barco

2 An Incessant Thirst: The Water Market in a Dried-Up Southern European Country, Spain, 1985–2022 15
Jesús Mirás Araujo and Nuria Rodríguez Martín

3 Between Private and Public: Water Supply Infrastructures in Portugal in the First Half of the Twentieth Century
An Attempted Comparison with Spain 35
Maria Ana Bernardo and Ana Cardoso de Matos

4 Public Health, Sanitary Movement and Public Water Supply Services in the Nineteenth Century: The Cases of England and Spain 60
Nuria Magaldi

5 Water Supply in the City of Cadiz in the Nineteenth Century: *The Cadiz Water Works Company Limited* 78
María Vázquez-Fariñas

6 The Improvement of the Water Supply Infrastructures in Antequera, 1818–1932 103
José Escalante-Jiménez, Mercedes Fernández-Paradas and Carlos Larrinaga

7 Foreign Companies, Public Intervention, and Ecological Crisis: Water in Seville, 1871–1959 121
Alberte Martínez-López

8 The New Supply of Potable Water in Malaga (Spain) in the 1920s: A Sustainable System? 144
Víctor Manuel Heredia-Flores

9 Water Supply in the El Aljarafe District (Seville Province, Spain): A Historical Perspective 162
Jesús Raúl Navarro-García

10 The "Price" of Water in Andalusia: An Approach to the Tariff Systems of the Domestic Water Supply Service 187
María José Vargas-Machuca Salido

11 Research on Wastewater Taxes: An Essential Contribution to Improving Environmental Management 210
Leticia Gallego-Valero, Encarnación Moral-Pajares and Isabel María Román-Sánchez

12 The Symbolic Configuration of Andalusian Rivers: Iconography and Artistic Values 225
Antonio Rafael Fernández-Paradas and Rubén Sánchez Guzmán

Index 245

Preface

The main objective of this book is to analyse the origin, development, and evolution of the public water supply service in Andalusia (1800–2020). Highly important issues have arisen in relation to the provision of this service which has acquired great relevance in recent years. In essence, the main objective is to contribute to furthering the knowledge and the analysis of the processes of municipalisation and/or privatisation of this service in Andalusia. Similarly, another aim of this work is to contribute to providing those responsible for local governments and administrations, on both a political and technical level, with a useful reflection and illustrative information on the use of municipalisation and/or privatisation as an instrument for reforming the local public sector.

To achieve this objective, we believe that the book should be based on several perspectives. First, the historical perspective is important in order to understand how the management of this public service has evolved over the last 200 years. To a large extent, this knowledge allows us to understand many of the current actions. Second, the economic perspective should be considered since it is useful to gain an idea of the level of efficiency of the drinking water supply service in Andalusia. Third, the legislative perspective illustrates how the regulatory framework of each era has determined the management and service provision models. And finally, the ecological and environmental perspective is highly important in the light of the New Water Culture and in the protection of the resource.

Within this context, the book is oriented towards the CHALLENGES of Andalusian society. Specifically, it focuses on the fifth challenge: Climate action, environment, resource efficiency and raw materials, which seeks a more efficient economy and society in the use of natural resources and water. Among the various lines of action of this challenge, the research focuses on environmental protection and the sustainable management of natural resources, water, biodiversity, and ecosystems, in order to provide knowledge and instruments that contribute to the management and protection of natural resources, enabling a sustainable balance to be achieved between the scarcity of resources and the current and future needs of Andalusian society and the economy. To achieve this, we will propose water management models that allow for an efficient and sustainable administration of water resources. In addition, we will provide knowledge and tools that enable effective decision-making and public commitment.

The nineteenth century saw the emergence of the "Modern Water System", which resulted in the model that is currently used in towns. Our intention is to study the process of how this new system was implemented: a universal, continuous, and permanent service, the introduction of a tariff system, the emergence of private companies, municipalisation processes, the emergence of public companies, etc.

The expansion and modernisation of municipal utilities and network industries requires the design and incorporation of new management and regulatory models. Western Europe was a pioneer in this respect, transplanting its public service management models to other countries. On the one hand, the French experience was notable for its decentralised pattern of public service at the municipal level, structured around private enterprise and with the introduction of competition through tendering processes and concession contracts. On the other hand, the British model stood out for its process of consolidation of the "service provider" in order to take advantage of economies of scale, privatising the agents that provided the service and creating a sophisticated system of economic regulation (Milward 2004 and 2007).

Spain has largely imitated the French model and the drinking water supply is based on public ownership by local administrations, but with the transfer of management to private companies. This system underwent a back-and-forth process, culminating in a system of concessions to both public and private companies (Matés-Barco 2013, 2014 and 2016). Subsequently, regulation has been used as a public policy technique. The direct control of services has been replaced, by way of concessions, with contracts with private companies operating on a monopoly basis.

Approaches to regulation are manifold and have generated much debate (Bel 2006). The academic literature, especially British and American publications, has addressed economic regulation from very different perspectives. Some authors have stressed the need for state intervention to smooth market failures, especially in the presence of monopolies or oligopolies. This control has been exercised through state regulation, so as to avoid the shortcomings of private agents and possible systemic risks in the economy. Others have shown the drawbacks of public interventionism (state, regional or municipal) arising from agency problems (Stiglitz 2009; Moss and Cisternino 2009).

In the case of Spain, studies can be found from the perspective of both administrative law and economics. There has been a tendency to reduce the role of the public sector, with the consequent promotion of privatisation, due to a desire to improve the functioning of public services. Terms such as nationalisation and monopoly have begun to be questioned and new concepts such as competition regulation and regulatory bodies have emerged. The continuous social, technical, and economic change promoting variations in management methods and formulas that adequately regulate these commercial activities is sought (Ariño, De la Cuétara, and Martínez 1997; Bel

2006; González-Gómez 2006; González-Gómez and García-Rubio 2008; Mirás and Piñeiro 2006; Matés-Barco 2017a, 2017b and 2017c).

This book analyses certain postulates of economic theory on the management of public services and, more specifically, on the management of the drinking water supply in cities. At the same time, it focuses on the case of Andalusia, analysing its historical evolution, the current situation of this public service and the existing management model in the Autonomous Region, using Cost-Benefit Analysis (CBA) or Cost-Efficiency Analysis (CEA) depending on the information available.

The editor and authors are researchers of recognised prestige, with extensive experience in this type of work and with a wide dissemination of their research. All of them have conducted studies that have been assessed by national accreditation agencies. Their work has been published in indexed journals and by prestigious publishers in Spain.

This book is a selection of works by prestigious professors from several Spanish universities (A Coruña, Cordoba, Granada, Jaen, Malaga) and the Spanish National Research Council (*Consejo Superior de Investigaciones Científicas*).

In the first chapter, Juan Manuel Matés-Barco (University of Jaen), analyses the management of the water supply in Andalusia between 1970 and 2000. He also proposes a possible explanation of the factors that have determined the management models. Within this context, he highlights the role played by European provisions, Spanish legislation, and the concept of public service in Europe.

Next, Jesús Mirás Araujo (University of A Coruña) and Nuria Rodríguez Martín (Complutense University of Madrid) study "An Incessant Thirst: The Water Market in a Dried-up Southern European Country, Spain, 1985–2022". The aim of this chapter is to analyse the transformations that have taken place in water regulation and management in Spain since the advent of democracy (especially since the approval of the 1985 Water Act), the map of uses and the main operators working in a sector that, despite the public nature of the service, has undergone a growing process of privatisation that in recent times seems to have moderated and even changed direction.

In the third chapter, Maria Ana Bernardo and Ana Cardoso de Matos (both from the University of Evora), analyse the history of the establishment of modern water supply infrastructures in Portugal in the first half of the twentieth century, focusing on the solutions adopted for the management of water supply networks and the legal and institutional framework that led to them. They also compare the Portuguese and Spanish cases, based on the same chronology and perspective of analysis.

Based on her research on public services, Nuria Magaldi (University of Cordoba) addresses "Public Health, Sanitary Movement and Public Water Supply Services in the Nineteenth Century: The Cases of England and Spain". The aim of her work is to analyse how the social hygiene movement was crucial to the development of the first public water supply mechanisms in Europe. In the nineteenth century, public health was largely based on the preventive dimension: its main objective was to preserve society from epidemic diseases and to provide it with better living conditions. To achieve this goal, the English sanitary movement (which spread throughout Europe, including Spain) strongly proposed to replace the traditional public fountains with a modern water supply system. The last part of the chapter, therefore, analyses how a modern public water supply service was created and managed in many municipalities in Spain, with a particular focus on the Andalusian region.

In chapter five, María Vázquez-Fariñas (University of Malaga), presents a case study after having carried out extensive research on "Water Supply in the City of Cadiz in the Nineteenth Century: *The Cadiz Water Works Company Limited*". Water supply has been one of the most important services in Spain since the mid-nineteenth century. Therefore, this chapter presents an analysis of the public water supply service in the city of Cadiz (Andalusia, southern Spain) during the nineteenth century. Specifically, it studies the water supply in the city, the nature of the management and supply of this resource and its main characteristics. In addition, this work examines the group of private companies that emerged in the city of Cadiz, and which contributed to the development of the sector. In particular, it analyses the role played by the London-based company *The Cadiz Water Works Company Limited* in the provision of drinking water management and supply services to the city of Cadiz. Thus, this chapter seeks to address the main characteristics of this company and the sector, and the most relevant problems that prevailed in order to determine how the water supply and management service was developed in the city.

The sixth chapter is written by José Escalante-Jiménez (University of Malaga), Mercedes Fernández-Paradas (University of Malaga) and Carlos Larrinaga (University of Granada), who have conducted the study "The Improvement of the Water Supply Infrastructures in Antequera, 1818–1932". This chapter addresses the main initiatives adopted by the local government of Antequera to improve the water supply between 1818 and 1932, in the context of the process of the modernisation of the municipal public services. These projects were characterised by a lack of continuity over time. The first, dating back to 1818, failed. The second, implemented between 1865 and 1867, resulted in a substantial improvement of the water supply. The third and most important project, developed between 1929 and 1932, was aimed at fully introducing the modern water supply system.

Next, Alberte Martínez-López (University of A Coruña) researches "Foreign Companies, Public Intervention, and Ecological Crisis: Water in Seville, 1871–1959". This chapter seeks to analyse, from the perspective of business history, the role played by British capital in the management of the water supply in Seville for almost seventy-five years through *The Seville Waterworks company*. Using primary business and public sources, it examines the origins of the company, its shareholding and organisational structure and business results, which were conditioned by demographics and rainfall, different economic circumstances, and public regulations, all within the framework of the evolution of the water sector in Spain and British investment abroad.

The eighth chapter, written by Víctor Manuel Heredia-Flores (University of Malaga), is titled "The New Supply of Potable Water in Malaga (Spain) in the 1920s: A Sustainable System?". The model of private management of the water supply service in the city of Malaga ended with its municipalisation in 1913. Serious problems with the quality and quantity of the supply continued until the 1920s. At that time, the engineer José Bores proposed a complete renovation of the system on new bases: a modern water supply without energy consumption, a new urban distribution network, a new sewerage network, a solution for industrial and agricultural use of the springs of Torremolinos, a hydrogeological study of the aquifer, sanitary control of the water and the establishment of an operation based on the use of meters. These aspects, especially the concern for the environmental preservation of the aquifer and its surroundings (creating a protection zone by planting trees) and the integral vision of the problem of supply and sanitation, made the Bores plan unquestionably modern and worthy of being studied in this book.

Jesús Raúl Navarro-García, a researcher at the School of Hispano-American Studies and the Spanish National Research Council (*Consejo Superior de Investigaciones Científicas*), is the author of the chapter titled "Water Supply in the El Aljarafe District (Seville Province, Spain): A Historical Perspective". To date, there have been very few studies on water supply and treatment in rural Spain, but this study analyses the situation in a district very close to Seville, the capital of Andalusia. Although this region is now practically part of the metropolitan area of the city, it was almost entirely rural until the middle of the last century. Until then, the water supply had been characterised by the use of its own resources, until population growth and the increased pressure of human activities exposed the failures of a system based on the use of springs and public and private wells. In response, external water resources were sought, and confederal management was promoted with the support of the Seville Provincial Council. This led to the creation of the current public company, ALJARAFESA, which manages water for 30 small and medium-sized municipalities with populations ranging from barely six hundred inhabitants to more than forty thousand, in a wide area with large pumping needs and many kilometres of pipes.

Next, María José Vargas-Machuca Salido (University of Jaen), presents the chapter entitled "The "Price" of Water in Andalusia: An Approach to the Tariff Systems of the Domestic Water Supply Service". From a legal point of view, in Spain, water as a good does not have a price, as it forms part of the public domain. Domestic consumers do not pay for the water they consume per se but for the services for its supply and treatment. The local authorities are responsible for setting these charges and their legal nature will determine the legal form of these considerations. Following the Water Framework Directive 2000/60/EC, water tariffs have become an important tool for achieving economic efficiency, environmental sustainability, and social equity. In this chapter, after defining the framework of competence for approving and setting the tariffs in Andalusia, a comparative approach is adopted to analyse the tariff systems for domestic water supply services in the eight provincial capitals of this region.

Leticia Gallego-Valero (University of Jaen), Encarnación Moral-Pajares (University of Jaen) and Isabel María Román-Sánchez (University of Almeria), present a study on "Research on Wastewater Taxes: An Essential Contribution to Improving Environmental Management". Wastewater taxes are an essential tool for the proper management of water resources, contributing to the sustainability of the planet. The bibliometric analysis carried out through WOS reveals the need for more scientific contributions on wastewater taxation, as there are insufficient results in this field of research for the period 2000–2020. A worldwide search yielded 144 results, representing less than 1.5% of the total research on all environmental taxes. Spain, the United States and China account for 50.68% of the total. This chapter is aimed at researchers in this field and seeks to enhance their understanding of the current research environment and to identify future trends.

Finally, Antonio Rafael Fernández-Paradas (University of Granada) and Rubén Sánchez Guzmán (Cultural Centre of the Madrid City Council), study "The Symbolic Configuration of Andalusian Rivers: Iconography and Artistic Values". The aim of this research is to identify, analyse and contextualise the phenomenon of the identification of Andalusian rivers with allegories. This is a national and international phenomenon, which developed a series of iconographic particularities in Andalusia. It is also a topic that extended into the twentieth century.

With all the above, this book is a novel work, covering a wide variety of subjects that have never been studied before, but with an important nexus: the relevance of the water supply service and its configuration in Spain and, more specifically, in the main Andalusian regions. Therefore, this book aspires to serve as a reference for all researchers, academics, teachers, students, and readers in general interested in the subject.

Juan Manuel Matés-Barco
María Vázquez-Fariñas

References

Ariño, G., J. M. De La Cuétara, and J. L. Martínez López-Muñiz. 1997. *El nuevo servicio* público. Madrid: Marcial Pons.

Bel, G. 2006. *Economía y política de la privatización local.* Madrid: Marcial Pons.

González-Gómez, F. 2006. "¿Está justificada la privatización de la gestión del agua en las ciudades? Teoría y evidencia a partir del criterio de eficiencia." *Ciudad y Territorio. Estudios Territoriales* XXXVIII (147): 139–57.

González-Gómez, F., and M. A. García-Rubio. 2008. "Efficiency in the Management of urban water services. What have we learned after four decades of research?" *Hacienda Pública Española / Revista de Economía Pública* 185: 39–67.

Matés-Barco, J. M. 2013. "La conquista del agua en Europa: los modelos de gestión (siglos XIX y XX)." *Agua y Territorio / Water and Landscape* 1: 21–9. https://dx.doi .org/10.17561/at.v1i1.1030.

Matés-Barco, J. M. 2014. "Las empresas concesionarias de servicios de abastecimiento de aguas potables en España (1840–1940)." *TST, Transportes, Servicios y Telecomunicaciones* 26: 58–89.

Matés-Barco, J. M. 2016. "La regulación del suministro de agua en España, siglos XIX y XX." *Revista de Historia Industrial* 61: 17–49.

Matés-Barco J. M. 2017a. "La regulación de los servicios públicos de gas y agua en Gran Bretaña (1800–1914)." In *Globalización, nacionalización y liberalización de la industria del gas en la Europa latina (siglos XIX-XXI)*, edited by I. Bartolomé-Rodríguez, M. Fernández-Paradas and J. Mirás-Araujo, 45–74. Madrid: Marcial Pons.

Matés-Barco, J. M. 2017b. "A regulação do abastecimento de água potável na Europa: Grã Bretanha e Espanha em perspectiva histórica / Regulation of drinking water supply in Europe: Great Britain and Spain in historical perspective." *Historia Económica & Historia de Empresas* 20(1): 9–50. https://doi.org/10.29182/hehe.v20i1.487.

Matés-Barco, J. M. 2017c. "El servicio público de abastecimiento de agua en España (siglos XIX y XX): El proceso de acumulación de competencias de los ayuntamientos." *Revista Brasileira de História & Ciências Sociais – RBHCS* 9(18): 36–57. https://doi .org/10.14295/rbhcs.v9i18.448.

Millward, R. 2004. "Regulation and ownership of public services in Europe: an historical perspective c.1830–1950." *Economia Pubblica* 34(2): 25–39.

Millward, R. 2007. "La distribution de l'eau dans les villes en Grande Bretagne aux XIXe et XXe siècles: le gouvernement municipal et le dilemme des compagnies privées." *Histoire, Economie, Société* 26(2): 111–28. https://doi.org/10.3917/hes.072.0111.

Mirás, J., and C. Piñeiro. 2006. "Tensions between Public and Private: Water Supply in a Northwestern Spanish City under the Franco." *Business and Economic History On-line* 4: 1–27.

Moss, D. A., and J. A. Cisternino (Eds.). 2009. *New Perspectives on Regulation*. Cambridge & Massachusetts: The Tobin Project.

Stiglitz, J. 2009. "Regulation and Failure." In *New Perspectives on Regulation*, edited by D. A. Moss and J. A. Cisternino, 11–24. Cambridge & Massachusetts: The Tobin Project.

Acknowledgements

This publication has benefited from the advice, support and help of many people. First, special mention should be made of the anonymous external reviewers for the suggestions and guidance they provided to improve this book.

Second, thanks are due to the Brill group for the warm response to the proposal for publishing this project. Particularly significant has been the help and attention received from Rosanna Woensdregt as the associate editor in charge of Brill's Biology and History of Science list, and manager of "Brill's Series in the History of the Environment". Her cooperation and support have been invaluable.

Thanks also to all the authors who have agreed to participate in this book, enriching it enormously with their contributions.

Finally, this research forms part of the results of the ERDF-University of Jaen-1381621 project (2021–2022) entitled "La gestión sostenible de los servicios públicos: Agua en Andalucía (1820–2020)", within the ERDF Operational Programme of Andalusia R+D+i (2014–2020) and which has been financed by the Regional Government of Andalusia and ERDF funds.

Figures and Tables

Figures

5.1 Remains of the Roman aqueduct at Cadiz 81
5.2 Emblem used by the company *The Cadiz Water Works Company* (*Limited*) 89
5.3 Advertisement of *The Cadiz Water Works Company* (*Limited*) in the *Guías de Cádiz* in the last third of the nineteenth century 96
7.1 Water contracts and consumption, in thousands of m^3, 1887–1901 126
7.2 Economic and financial profitability, 1887–1956 127
7.3 Exchange rate pta/pound, in pesetas per pound, 1887–1959 128
7.4 Rainfall in Seville, 1880–1950, in mm 129
8.1 Official inauguration of *Torremolinos Waters* pipeline, 1876 152
8.2 Bridge over the Guadalhorce River 156
8.3 Olletas water reservoir under construction 157
9.1 Women with their earthenware *cántaros* after collecting water from the *El Rajao* well (Benacazón) 170
9.2 *El Rajao* well (Benacazón). Private wells such as this helped to provide water to residents in the locality 171
9.3 Water carriers in Carrión de los Céspedes, with a specially prepared wooden barrel on a cart drawn by a mule 173
9.4 Eighteenth-century baroque fountain in Aznalcázar. An important example of El Aljarafe's water heritage 180
10.1 Price of the fixed amount for domestic water supply services (20 mm calibre meter) (Euros per month) 203
11.1 Environmental taxes by area as a percentage of total results for environmental taxes 214
11.2 Global and Spanish scientific contribution to research on environmental taxes by area (2000–2020) 215
11.3 Trends in the research on wastewater taxes 215
11.4 Main areas of study in wastewater taxes research (percentage) 216
11.5 Top countries in wastewater taxes research 217
11.6 Main institutions in wastewater taxes research 220
12.1 Anonymous. Lintel of the rivers. Ana and Albarregas rivers 229
12.2 Gaspar Salcedo de Aguirre. Map of the kingdom of Jaen 233
12.3 Pedro Anastasio Bocanegra. Allegory of the River Darro. Circa 1676–1680 234
12.4 Anonymous. Allegory of the River Guadalquivir. Circa 1770 235
12.5 Ángel María Barcía Pavón. Allegory of Cordoba and the Guadalquivir. 1892 238

Tables

1.1 Forms of management of local public services 7

5.1 Prices for the sale of water to individuals during the first 50 years of the concession 85

5.2 Capital necessary for executing the water supply project, 1870 86

5.3 Prices for the sale of water in perpetuity 88

5.4 Tariffs of *The Cadiz Water Works Company, Ltd.* 93

5.5 Tariffs of *The Cadiz Water Works Company, Ltd.* for monthly leasing and repairs of the meter 94

7.1 Population of Seville and inter-census growth 1860–1960 122

7.2 Shareholders in 1891 125

9.1 Localities in the El Aljarafe district (Seville province) distributed by number of inhabitants in 1940, 1970, 2001 and 2021 168

10.1 Water supply service: managing entities, by-laws and updates of tariffs 201

10.2 Tariff structure of the supply of water for domestic use in the Andalusian provincial capitals 202

10.3 Bonuses to the tariffs of the supply of water for domestic use 204

11.1 Journals and relevant indexes in wastewater taxes research 218

11.2 Main authors in wastewater taxes research 219

Notes on Contributors

Juan Manuel Matés-Barco

Professor of History and Economic Institutions at the University of Jaen. Graduated in Geography and History from the University of Zaragoza and Doctor in History from the University of Granada. He has four six-year research periods with research experience in the field of public services and the economic evolution of contemporary Spain. He is a collaborator of several international magazines. He has made stays at the Università degli Studi di Firenze, at the European University Institute (Italy) and at the Université Michel de Montaigne Bordeaux 3. He is head Researcher of the Group of Historical Studies on the Company (GEHESE-UJA), Researcher at the Permanent Seminar on Water, Territory, and the Environment: Public Policies and Citizen Participation. Coordinator of various research projects, and Director of the journal *Agua y Territorio / Water and Landscape* (AYT/WAL).

María Vázquez-Fariñas

Lecturer in History and Economic Institutions Area (Department of Economic Theory and History) at the University of Malaga (Spain). Ph.D. in Social and Legal Sciences and Degree in Business Administration and Management from the University of Cadiz. She is actively involved as a component of the Research Group of Historical Studies on the Enterprise (GEHESE-UJA). She has research stays in Spanish and foreign centres of recognized prestige, such as the School of Hispano American Studies (CSIC, Seville), the University Michel de Montaigne Bordeaux 3, and the University of Cadiz. Her general lines of research focused on the economic and business history of nineteenth-century Cadiz (Andalusia, Spain), and the development of gas, water, and public services in contemporary Spain. She is author of many works on these subjects published in national and international journals and publishing houses, such as Marcial Pons, Routledge, Peter Lang, and Palgrave. Also, she participates in several research and teaching innovation projects and part of the Editorial Board of the Journal *Agua y Territorio / Water and Landscape* (AYT/WAL) from 2018 to the present.

Maria Ana Bernardo

Assistant professor at the University of Evora and researcher at the CIDEHUS/EU research centre. She holds a Ph.D. in History from the University of Evora. Her research interests are centred on Urban History, Sociability Practices, History of Leisure and Tourism, Political Elites and Social Stratification. She has participated in some projects funded by the Science and Technology

Foundation of Portugal, including: "Networked Cities: Urban Infrastructures in Portugal 1850–1950" and "Cultural life in provincial towns. Public space, sociability, and representations (1840–1926)". She is author of many works on these subjects published in national and international journals and publishing houses.

Ana Cardoso de Matos

Professor at the University of Evora and researcher at the CIDEHUS/UE research centre. Since 2007, she has been the coordinator at the University of Evora of the *Erasmus Mundus Master TPTI – Techniques, patrimoines, territoires de l'industrie*, and has taught at nine different international universities. She is member of the Comité d'historie de l'electricité et de l'enérgie de la Fondation EDF since 2012, member of the board of the International Railways History Association (IRHA) since 2013, and member of the Iberian Railway History Association since 2016. She is part of the Editorial Board of *TST – Transportes, Servicios y Telecomunicaciones*, and *Journal of Energy History (JEHRHE)*, and member of the Scientific Committee of *Patrimonio Industriale AIPAI* (Italy); *e-Phasistos. Revue d'histoire des techniques*; *Midas-Museums* and *Interdisciplinary Studies*. She is part of several national and international projects and publishes regularly in both peer-reviewed journals and books from reputable publishers in the scientific community.

José Escalante-Jiménez

Doctor in History from the University of Malaga. Director of the Royal Academy of Antequera. He is official Chronicler of the City of Antequera, Director of the Municipal Historical Archive of Antequera from 2005 to 2012, Archivist of the Municipal Historical Archive of Antequera, full member of the Illustrious Andalusian Society of Legal Historical Studies of Cordoba, and Academic of the Royal Academy of Fine Arts of San Telmo in Malaga.

Antonio Rafael Fernández-Paradas

Doctor in Art History from the University of Malaga, with the doctoral thesis entitled "Historiography and methodologies of the History of furniture in Spain (1872–2011). A state of affairs". Graduated in History of Art and Bachelor of Documentation from the University of Granada. Master in Expertise and Appraisal of Antiques and works of art from the University of Alcala de Henares. He is currently a Contract Professor at the University of Granada, where he teaches in the Department of Didactics of Social Sciences of the Faculty of Educational Sciences, and teaches the Master's Degree in Art and Advertising at the University of Vigo.

Mercedes Fernández-Paradas
Doctor and Head Professor of Contemporary History at the University of Malaga. Specialised in Economic History and public services, especially business history, gas and electricity. She is principal researcher of the Research and Innovation Project of Excellence of the Government of Spain "La industria del gas en España: desarrollo y trayectorias regionales (1842–2008)". She has carried out several research stays, including at the London School of Economics and Political Science. She has won several awards, including the Premio Extraordinario de Doctorado en Filosofía y Letras [Extraordinary Doctorate Award in Philosophy and Arts] for the History Department of the University of Malaga.

Leticia Gallego-Valero
Ph.D. in Economic and Business Sciences from the University of Jaen, lecturer of Applied Economics and member of the "Economía Aplicada Jaén" research group. She has participated in different research projects financed by provincial and regional institutions, having developed, in recent years, intense study and analysis work in the field of taxes on wastewater and the management of municipal wastewater treatment services. The analysis and study of different economic and fiscal aspects in the environmental field are the focus of her research, published in international journals.

Víctor Manuel Heredia-Flores
Associate Lecturer in Economic History at the University of Malaga. Graduate in Contemporary History and Ph.D. from the University of Malaga with the thesis entitled "Private management and municipalisation in the supply to the city of Malaga. The Torremolinos Water Business (1860–1930)". He is a member of the project "Historical Statistics of Andalusia" of the Institute of Statistics of Andalusia. His research experience is in fields such as the history of education in Andalusia, the reconstruction of regional historical statistical series and industrial history and public services.

Carlos Larrinaga
Reader in Economic History at the University of Granada (Andalusia, Spain). His research is focused on the history of tourism, railways in the nineteenth century and the service sector. He has led four interdisciplinary projects on the history of tourism in the twentieth century, funded by the Spanish Ministry of Science, Innovation and Universities and the ERDF funds. He has undertaken research in several stays at Bordeaux-Montaigne University or at Aberystwyth University, among other.

Nuria Magaldi

Doctor of Law with European mention and extraordinary prize (University of Barcelona, 2010). She graduated in Law (2001) and in Political and Administration Sciences (2006). LL.M. (Greifswald University, 2001–2002). Diploma of the European Academy of Public Law (Greece, 2003). She was a fellow of the Generalitat de Catalunya (2003–2006) and assistant professor (2006–2008) in the Department of Administrative Law at the University of Barcelona. As of 2009, she joined the Administrative Law Area of the Faculty of Law and Economic and Business Sciences of the University of Cordoba, where she is currently a Full Professor. She has carried out research stays in Cordoba, Florence, Frankfurt, Heidelberg and at the Universidad Externado de Bogotá. She has published three monographs and various articles and book chapters and has done translations from German and Italian. She has two six-year research periods.

Alberte Martínez-López

Professor of Economic History at the University of A Coruña, where he coordinates the Business History Study Group. He has carried out research and teaching stays at universities in Europe and America. His lines of research are business history, utilities (transport, energy, water), foreign investment, and livestock. Related to these issues, he has published several articles in prestigious publishers (Routledge, Peter Lang, Marcial Pons, Síntesis, Sílex), and in high-impact Spanish (*Ayer*, *Historia Agraria*, *Revista de Historia Industrial*, *Asclepio*, *Investigaciones de Historia Económica*) and international journals (*Business History*, *Journal of Urban History*, *Journal of Urban Technology*, *Continuity and Change*, *The Journal of Transport History*). He has coordinated several research projects and contracts with companies and institutions.

Jesús Mirás Araujo

Senior Lecturer of Economic History at the University of A Coruña. B.Sc. Economics (University of Santiago de Compostela), M.Sc. Economics (University of A Coruña), and Ph.D. Economics (University of A Coruña). He holds three six-year research periods. His main research interests are on urban history, business history and public utilities (gas, electricity, water, transport), and port and maritime history. He is author and editor of several books about these topics, chapters in Peter Lang, Routledge or Presses de l'Université Paris-Sorbonne, and articles in *Business History*, *Continuity and Change*, *Urban History* or *Journal of Urban History*. He has participated in several research projects, and also in contracts with companies. He is Associate Editor of the *International Journal of Economics and Business Modelling*, and previously member of the Editorial Board of *TST. Transportes, Servicios y Telecomunicaciones*.

Encarnación Moral-Pajares
Ph.D. in Economics and Business at the University of Malaga (Spain). Teacher in the Department of Applied Economics of the University of Jaen (Spain). The analysis and study of different economic aspects related to urban wastewater are the focus of her latest research work, published in international journals. She has been the main person in charge of different research projects financed by national institutions and visiting professor at the Autonomous University of Lisbon in Portugal and at the Hochschule Furtwangen University in Germany. Arco Iris Award 2014, José Luis Sampedro Award for the best communication presented at the XVI World Economy Meeting held in Cadiz on June 2014, UNICAJA 2007 Award for Agricultural Research; Essay Prize Jaen 2001. She is Co-director of the EXTENDA Chair of Internationalization at the University of Jaen. She has been Vice-Dean of the Faculty of Social and Legal Sciences at the University of Jaen and Director of the Secretariat for Scholarships, Aid and Student Services at the University of Jaen.

Jesús Raúl Navarro-García
Scientific researcher for the Higher Council for Scientific Research (CSIC) at the History Institute's Department of Art and Heritage in Madrid. He holds a B.A. in Geography and History (Universities of Zaragoza and Seville), an M.A. in Landscape Protection, Management and Planning from the University of Seville, and a Ph.D. in the History of America from the University of Seville. He has been the director of the School of Hispano-American Studies (EEHA, CSIC) in the city of Seville for almost ten years. He is editor and co-founder of the journal *Agua y Territorio / Water and Landscape* (*AYT/WAL*) and also of the EEHA's Permanent Workshop on Water, Territory, and the Environment. He is the lead researcher of the Nature, Territory, and Cultural Imaginaries research group (CSIC) and a member of the Territorial Structures and Systems research group with investigators from Seville and Pablo de Olavide universities. He has five six-year research periods and five *quinquenios*. His research has focused on the colonial policy of Spanish liberalism, water supply and treatment in rural Spain, and the historical landscape in thermal villages.

Nuria Rodríguez Martín
Lecturer in Contemporary History at the Complutense University of Madrid (Spain). Ph.D. in History from the Complutense University in 2013. She has been the recipient of several fellowships, including a Post-Doctoral Fellowship at Malaga University (2019–2020), and a Juan de la Cierva-Fellowship in the Department of Modern History at the University of the Basque Country (2016–2017). She also holds a B.A. in Journalism (1995) and M.A. in Mass Communication, Information and Propaganda from the Complutense

University of Madrid (1996). Her broad areas of research interest include urban history and public utilities in the nineteenth and twentieth centuries Spain. She has published extensively about the history of water, gas, and electricity supply within the Spanish cities during the nineteenth and twentieth centuries. She is the author of two books and 15 scholarly articles.

Isabel María Román-Sánchez

Ph.D. in Economic and Business Sciences from the University of Almeria (2004), professor of Applied Economics, member of the research group "Environmental Economics and Natural Resources", belongs to CIESOL (Centre for Research in Energy Solar), has been a collaborating researcher in different research projects of the National Plan and the Junta de Andalucía, related to the purification and regeneration of wastewater. She has developed extensive research work in the field of water and environmental taxes and has published in numerous national and international journals. She has been Vice-Rector for Infrastructures, Campus and Sustainability at the University of Almeria, Director of the Social Affairs Secretariat and is currently coordinator of the Master's Degree in Accounts Auditing at the University of Almeria.

Rubén Sánchez Guzmán

Degree in Art History from the Autonomous University of Madrid. He currently works as a professor of Art History and Knowing Madrid at the Cultural Centres of the Madrid City Council (Educo-Activities). He has been a teacher of the I Master in Spanish Baroque Sculpture. Since the golden centuries the information society and social networks taught by the International University of Andalusia and Expert Appraiser in Antiques and Works of Art. Art Department. Lamas Bolaño Madrid Auction Room. Performing the functions of appraisal, cataloguing and personalized advice.

María José Vargas-Machuca Salido

Graduate in Economic and Business Sciences from the University of Navarra and Ph.D. from the University of Jaen. Lecturer in the Department of Economics at the University of Jaen, where she teaches the subject of the Spanish Financial System in various degrees. She belongs to various research groups and teams related to her area of knowledge. Her research has focused on the history of the financial system, especially at the local level, a subject on which she has published several articles in specialized journals and chapters in collective works published by renowned national and international publishers.

CHAPTER 1

Water Supply Management in Andalusia

Juan Manuel Matés-Barco

Abstract

This chapter seeks to analyse the management of the drinking water supply in Andalusia between 1970 and 2000. It also offers a possible explanation of the factors that have determined the management models. Within this context, particularly noteworthy is the role played by the European provisions, the Spanish legislation, and the public service concept existing in Europe.

Keywords

Andalusia – companies – management – public service – water supply

1 Introduction

Supplying drinking water to populations is an activity that is highly intensive in capital, with a very high degree of fixed costs and in which the duplicity of networks generates a high level of inefficiency. The public services related to water are closely linked to economies of scale and take the form of a natural monopoly (Núñez 2005 and 2018a). The economic literature on this aspect provides sufficient reasons to explain the absence of competition between companies (Matés-Barco 2006, 2018a and 2021b). Over time, there has been a clear evolution in the forms of management, giving rise to what could be described as a "return trip": the prevalence of private companies in the nineteenth century, a tendency towards direct management by the local governments from 1920 and a return to private management from the final decades of the twentieth century (Núñez 2018b; Matés-Barco 2021a; Cuervo 1986, 331–40).

 | DOI:10.1163/9789004679634_002

This chapter seeks to analyse the management of the drinking water supply in Andalusia between 1970 and 2000. It also offers a possible explanation of the factors that have determined the management models. Within this context, particularly noteworthy is the role played by the European provisions, the Spanish legislation, and the public service concept existing in Europe (Matés-Barco 2018a).

After the introduction, the second section briefly describes the debate existing on public services. The third section summarises the situation in Andalusia as a region that has suffered a delay in terms of the implementation of the drinking water supply service. Next, in the fourth section, the characteristics that have predominated in this sector between 1970 and 2000 are examined. Finally, the chapter ends with some brief conclusions and a short bibliography.

2 The Debate on Public Services

The operating problems of public services have generated a broad academic debate. Both the British and American literature have examined the issue, as has the European administrative law and, more specifically, that of Spain (Ponzellini and Treu 1994). In recent years, a trend has emerged supporting the reduction in size of the public sector with the consequent promotion of privatisation, due to a desire to improve its operations. In Spain, the operations of a good part of the state public services (electricity system, postal service, RENFE (rail service) or local services (water supply, abattoirs, markets), remain unaltered by the rapid technological and social evolution that have taken place in recent decades and are still anchored in the old structure inherited from the mid-twentieth century (Matés-Barco 2019b). The deficiencies observed are a result of the application of traditional formulas (nationalisation and monopoly) (Castro-Valdivia, Matés-Barco, and Vázquez-Fariñas 2020).

Establishing a clear distinction between the public and private is complex and it is difficult to determine where one ends and the other begins. A look back in time enables us to observe that economic crises have increased the trend towards the public management of certain services. On the contrary, in times of stability and peace, the balance has swung towards free trade and private management. However, continual social change promotes the constant mutability of the management concept. Contemporary societies are increasingly more complex, and, therefore, public services tend towards ramification and a larger size.

In the 1920s and 1930s, a major debate arose regarding their efficiency in the English-speaking world. The problem began to emerge again from 1980, with British privatisation and American deregulation[1] (Martínez 1997, 195) and today, the issue has been reopened once again. The public control of certain activities has been justified by their social importance. The establishment of a monopoly and exclusive ownership by the state has been common practice in the management of economic public services, although indirect forms of management have also been implemented. This approach has given rise to comparisons between public service and nationalisation. In recent decades, there has been a questioning of the excess level of interventionism adopted after the Second World War and a trend towards deregulation has been observed, particularly visible in the United States and Great Britain. This trend has given rise to a discussion on the traditional concept of public service born in the European continent with a shift towards the notion of the "regulated activity of general interest", typical of the English-speaking world (Ariño 1993 and 1997; Roversi 1996; Matés-Barco 2018c).

Several solutions have been implemented. The first is the "English-style" privatisation, which countries such as Chile have adopted with an acceptable level of success. This formula seeks to make the most of the capital and the thrust of the private initiative and to reduce excessive Statism. Second, "American-style" deregulation is based on a free competition regime among the private agents. Air transport has been the "star sector" in which this formula has been most broadly applied. The third solution can be referred to as the "French-style defence" of the traditional values of the public service. Finally, other dispersed initiatives include the direct and indirect management of public services ("Spanish-style privatisation") or the rigid economic approaches practised in Japan to address the sustainability of healthcare and social security for an increasingly ageing population.

The concept of public service as an essential element of the Welfare State has been questioned in recent decades. Its exclusive and monopolistic nature, which, in Spain has been justified with article 128.2 of the Constitution, has entirely lost its credibility. For some, the only exceptions were the old revenue-producing monopolies (the state monopolisation of salt, gunpowder, matches, tobacco, oil...), which gradually disappeared with the progressive application of the principles of economic liberalism. However, this does not mean that the "classic" public service was without its virtues. It enabled the access of a good

1 The expression "deregulation" essentially refers to a "reduction in regulations and the public control of activities".

number of citizens to a series of services which, otherwise, would only have been available to privileged minorities; it generated healthcare, social and educational infrastructures, which have facilitated the access to modernity; and promoted the satisfaction of primary vital needs. However, both Europe and Spain have experienced substantial changes that warrant the development of new public service models.

3 Andalusia: Scarcity and Delays

Drinking water supply services did not seem to attract the attention of investors in Andalusia at the end of the nineteenth century and the first half of the twentieth century. The data indicate that it was a sector with a significant presence of foreign investment and in which the local initiatives were marked by insecurity (Castro-Valdivia and Matés-Barco 2020). In the mid-twentieth century, together with Extremadura, Andalusia was one of the Spanish regions with the lowest indexes of municipalities with a domestic supply of drinking water. Specifically, in the province of Seville, just 30.39% of the municipalities had a water supply, only exceeded by Cordoba and Cadiz with 34.66% and 30.95% respectively. Nevertheless, Seville had the highest index of private companies engaged in the supply of water in the whole of the region. Of a total of 35 entities operating in the water supply business, a little over half of them belonged to the private sector (54.28%), as opposed to 30% in Cadiz and 28% in Cordoba. In the province of Jaen, 27.7% of the companies engaged in the supply of water were private; however, this figure is not particularly significant as only 1.5% of the municipalities had some type of domestic connection or supply (Matés-Barco 1997 and 2002).

After analysing the water supply in Andalusian municipalities, according to the number of inhabitants based on the statistics of 1950, we can observe the disproportion existing between the large and small towns. In the latter, those with less than five thousand inhabitants, the highest index of municipalities with a water supply was just 18.33%, corresponding to the province of Seville. If we examine poor provinces, the number of water supply companies was zero, as in the case of Almeria, Cadiz, and Granada. In other provinces, such as Huelva and Jaen, this percentage was around 6% and in Malaga and Cordoba 14.66% and 14.70% respectively (Matés-Barco 2019a and 2019b). The data confirm an undeniable fact: the supply of water, until after the 1950s, was an eminently urban phenomenon, disconnected from the rural world and provided to municipalities with more than 50,000 inhabitants (Heredia-Flores 2013).

4 Water Management in Andalusia in the Age of Democracy

With the installation of the democratic regime in Spain and the creation of the State of the Autonomous Regions following the Constitution of 1978, certain changes in water management progressively took place. Here, we should describe the role played by each of the agents involved in this task, namely the State, the regional governments, and local governments. The State coordinates the administrative tasks through the Ministry of the Environment on which different bodies depend, such as the General Directorate of Hydraulic Works and Water Quality, the Hydrographical Confederations – Guadalquivir, Guadiana, Segura, and Sur, in the case of Andalusia –, and certain state-owned companies: *Hidroguadiana*, *Aguas de la Cuenca del Guadalquivir* (*Aquavir*), *Aguas de la Cuenca del Segura* (*Acsegura*) and *Aguas de la Cuenca del Sur* (*Acusar*). The Hydrographical Confederations or basin organisations are bodies governed by public law with their own legal personality. They are attached to the Ministry of the Environment and have full functional autonomy, in accordance with the Water Law of 1985 (Matés-Barco 2004 and 2006).

Water management is a complex issue due to the many agents, objectives and activities involved in its regulation (Matés-Barco 2018b). Aspects such as agriculture, urban consumption, healthcare, fishing, energy generation or sport and leisure form a wide and heterogeneous range of activities which are necessary to reconcile. In Spain, after a complex and changing evolution, the State declared water a public domain so as to ensure the respect of the management unit in the basin and the water systems and water cycle. In the case of the Hydrographical Basins, it is worth mentioning the inter-community basins (territory spanning several Autonomous Regions) and the intra-community basins (belonging to just one region). In turn, the public management of water seeks to make territorial planning compatible with the conservation and protection of the environment and with the restoration of natural spaces which have been affected by large hydraulic works. On the other hand, a comprehensive management of water saving is encouraged which seeks to combine a coordinated decentralisation with the participation of the users (Martín et al. 2006a).

The water policy of the European Union forms part of the Water Framework Directive issued by the European Parliament in 2000. This regulation establishes lines of action on an EU level with respect to water policy. More specifically, it places particular emphasis on the protection of inland surface waters, groundwater, transitional waters, and coastal water. It also promotes tasks that guarantee urban supply and sustainable use and drives actions aimed at reducing pollution and mitigating the effects of floods and droughts. The European community regulations make special reference to the need to regulate the treatment of urban wastewater, stating that all towns of more than 2,000 inhabitants must establish a wastewater collection network and a wastewater treatment plant.

The State, as a guarantor of the water public domain, is responsible for water planning and infrastructure projects. It is also responsible for authorising and supervising the public water domain in inter-community water basins. Furthermore, it grants concessions and adopts measures to ensure the compliance with the agreements and international conventions. Other exclusive competencies of the State refer to aspects that affect several Autonomous Regions, such as the authorisation of the construction of electrical facilities, energy transportation or public works of general interest. A similar situation arises when waters run through more than one region, given that the State is responsible for the legislation, planning and concession of resources and water uses.

The regional governments, meanwhile, are responsible for the elaboration of projects and the construction and exploitation of hydraulic infrastructures, canals, and irrigation systems exclusively within their territory. They are also responsible for mineral and thermal waters. This is the case of the regional government of Andalusia, which has assumed the exclusive responsibility for water resources and their exploitation, canals and irrigation systems and their planning and concession whenever the waters run entirely through its territory. These competencies are shared between the different regional departments: Public Works and Transport, Environment, Agriculture and Fishing and, finally, Health.

There are other subordinate bodies, such as the General Secretariat for Water, which, in turn, depends on the Water Institute of Andalusia, which defines the water policy and establishes guidelines for research. Along these lines, in 1994, the Centre for the Research, Promotion and Application of New Water Technologies (CENTA) was created with the objective of contributing to the development of studies promoted in coordination with universities, the corporate world and the Regional Government of Andalusia. Other similar bodies are the Directorate General for Irrigation and Structures, the Directorate General for Prevention and Environmental Quality and the Directorate General for Public Health and Participation. Since 2004, the Andalusian Water Agency has undertaken the functions which, until then, had been assigned to the Secretariat for Water. The essential tasks of this organisation focus on the investments for elaborating infrastructure projects and works. The principal investment activity is focused on the supply of water (51.34%), followed by sanitation and wastewater treatment (32.40%) and the channelling and protection of riverbanks (13.34%).

Another new feature has been the constitution of public companies related to water management. The *Empresa Pública Gestión de Infraestructuras, S.A.* (GIASA) is attached to the Department of Public Works and Transport and its activity focuses on developing projects and the construction of hydraulic infrastructures. The *Empresa Pública para el Desarrollo Agrario y Pesquero de Andalucía, S.A.* (DAP) is attached to the Department of Agriculture and Fishing and its objective is the modernisation and improvement of equipment and facilities to promote a more rational use of water and the management of water resources, the installation of treatment plants and water treatment.

Throughout history, the local governments have been responsible for aspects related to water supply. More recently, these competencies have been extended to sewerage systems and wastewater treatment. The management of these public services has evolved considerably over the last few decades. The most recent legislation distinguishes between "direct" and "indirect" management. In the former case, the services are provided "by the local council or an autonomous local body", although the possibility of the existence of a public company which is wholly owned by the local government is also contemplated. In the latter case, "indirect management", the law considers different formulas, such as the "concession", "interested management", "leasing" "concerted management" and the creation of "commercial or mixed companies" owned by several agents which may include the local government (Table 1.1). In any event, the tariffs must always be approved by the Regional Government.

The data for the year 2000 show that most of the local councils of Andalusia directly managed the water supply service. Of the 770 municipalities existing in the Autonomous Region, this was the system chosen by 561, that is,72.8%. However, the predominant formula in the municipalities with the

TABLE 1.1 Forms of management of local public services

Direct management	Indirect management
Local council	Concession
Autonomous body	Interested management
Public company	Concerted management
	Leasing
	Mixed companies

SOURCE: OWN ELABORATION

largest populations was the concession. There were many towns in this situation (53%) that provided the service to more than 4 million inhabitants (75% of the urban population and 58% of the total population of Andalusia). Direct management, on the other hand, prevailed in 47% of the large municipalities, although only representing 25% of the urban population.

The vast majority of municipalities with less than 10,000 inhabitants opted for direct management (88.7%) and of those with a population above this threshold, only 11.2% preferred this formula. The data show that the large cities leaned towards the assignment of the service, while the smaller or rural towns chose direct management. This pattern may also be observed in a provincial analysis and the differences existing between one and another were due to the minimum variations in the structure of their settlements.

A trend towards privatisation can be observed with its respective forms in the provision of local public services: drinking water supply, sanitation, cleaning, collection, and treatment of urban solid waste, etc. On the contrary, direct management is used in services more recently incorporated into the municipal competencies, particularly aspects such as the environment, the provision of social, cultural or sports services, the creation of employment and all elements related to the development of the local economy.

Several features characterise and condition the provision of the drinking water supply service and the undertaking of an adequate sanitation of wastewater. First, the structure of the population, its distribution, and places of settlement; and second, the size and quality of the infrastructures, particularly the state of conservation of the capture and distribution network. The population distribution in Andalusia is clearly uneven. Most of the population (62.3%) resides in a small number of municipalities (7.8%), while only 4% of the population lives in 41.6% of the municipalities.

Water is mostly collected from wells, boreholes or underground galleries and the connections made directly with the springs are relevant, particularly in the provinces of Malaga, Granada and Jaen (Heredia-Flores 2018). No full assessment has been made of the state of the water collection, although, in general terms, the regional government classes it as acceptable. On the contrary, the conveyance network has evident deficiencies in some provinces, of which we can highlight Cadiz (66.62%), Cordoba (51.99%), Seville (45.60%) and Granada (38.04%). The average for Andalusia is several points below that of Spain as a whole and only three Andalusian provinces positively exceed the national average, indicating the need for investments and improvements in this field.

The volume of water that can be stored in tanks is lower than the national average (0.44 m^3 as opposed to 0.60), as is the regulation capacity (1.56 days as opposed to 1.89). These data also reveal the level of delay existing in these types of infrastructure. Meanwhile, almost the entire population and all of the municipalities receive purified water. This is an improvement with respect to the Spanish national average by several points. The public distribution network is the system implemented across all of the Andalusian provinces, reaching 99.66% of the population. The management of the network is undertaken directly by the local council or through a municipal public company. In this section, private management is almost irrelevant. The sanitation network is extensive and reaches 98.9% of the population, exceeding the Spanish average by several points (92.5%). Private management of sanitation is very rare in the region of Andalusia and the majority of the municipalities prefer a management that is direct or conducted through a public company. Jaen and Granada are the provinces with the sanitation network in the worst state and Almeria has a considerable deficit in terms of the kilometres of the network.

The installation of infrastructures necessary for the treatment of wastewater is a task that is still pending in the Autonomous Region of Andalusia, particularly in provinces such as Seville, Cordoba, Cadiz, Granada, and Jaen. The conventional treatment plants are those most used and the "lagoon" system is the most adopted in the provinces of Seville, Jaen, and Almeria. At the end of the twentieth century and beginning of the twenty-first century, the number of wastewater treatment plants rose considerably, but a large percentage of the population (over 50%) were still left without this service (Martín et al. 2006b).

Over the last few decades of the twentieth century, the local councils of Andalusia adopted the indirect management of the water supply service through public and private companies. This type of company managed the domestic supply of almost 40% of the municipalities in Andalusia and 80% of the population. Public, private and mixed companies, in this order, were the formulas used by the local councils to undertake the management of the service.

In the small municipalities, the local council directly managed the provision of the water supply service. The medium-sized municipalities usually recurred to the public company, while the larger municipalities usually opted for the solution of the mixed or private company. Therefore, the externalisation of the service was more frequent in medium and large municipalities. Only the small towns favoured direct management. The models most used by the medium-sized and large municipalities were those based on a management through public and private companies. At the beginning of the twenty-first century, in the whole of Spain, the population supplied by private companies accounted for 37%, while in Andalusia this was 23%. The opposite was the case with direct management. In Spain, only 7% of the population was supplied by this system, while in Andalusia this figure was 19%.

The water supply sector is highly dynamic and has undergone frequent changes in the ownership of the management due to processes of absorption, mergers, or sales of the shareholding (Fernández-Paradas and Matés-Barco 2022). Furthermore, we can observe a progressive consolidation of a few large companies managing the service either exclusively or through participation in a mixed company. There are several important private groups that operate in Andalusia. First, we can refer to *Dragados y Construcciones* which, through *Urbaser*, manages the service in municipalities such as Luisiana, Isla Mayor and Lentejuela and participates in *Aguas del Huesna*. Second, *Ferrovial* created the company *Ferroser* to develop activities in the sector in towns such as Estepona, Velez Blanco and Ubeda. Third, the French group *SAUR* is one of the most long-standing in the sector with extensive experience in several continents. It participates in the company *Gestión y Técnicas del Agua* (Gestagua), which manages the service in towns such as Fuengirola, Albox and Montoro.

Finally, we should mention the two companies with the greatest presence in terms of the provision of the service in the Andalusian municipalities: *Aqualia* and *Aquagest*, which belong to different corporate entities. The first, *Aqualia*, forming part of the business group *Fomento de Construcciones y Contratas* (*FCC*), manages the service in cities such as Jaen or Almeria. Second, *Aquagest* is a company dependent on *Aguas de Barcelona*, which is one of the business groups with the longest tradition in the sector. Its presence in Andalusia has been secured through the obtaining of direct concessions for the management of the service, such as the case of towns such as Marbella, San Fernando, or Roquetas de Mar. In other cases, it participates in mixed companies such as *Emasagra* and *Aguasvira* in Granada, or *Astosam* in Torremolinos.

Other groups and companies are less prominent in the sector but are also fairly significant. This is the case of companies such as *Abengoa*, *Endesa*, *Iberdrola*, *OHL*, *Hidrocartera* (*Unicaja* and *Caja de Granada*), or the German group *RWE Thames Water*. In general, they participate in the sector through intermediary companies or affiliates such as *Aguas de Herrera*, *Aguas de Guadix*, *Ondagua*, *CRIDESA*, *Inima*, etc.

5 Conclusions

The insecurities and fluctuations experienced by the drinking water supply sector in Andalusia, particularly in the period 1850–1950, influenced the subsequent development of this public service. The whole of this period was characterised by the budgetary, civil service, technical and financial incapacity of the municipalities. Difficulties were encountered to establish the water supply, first due to the technical exogenous resistances and second, because it was too heavy a burden for the slight shoulders of the local councils. This aspect reflected their tremendous weakness when attempting to exercise municipal power, even in large industrial cities where they were subordinated to the private interests of either owners or service companies. As a result, the municipal services ended up in the hands of private companies and concessions were granted to resolve this supply problem.

The difficulties encountered by many of these private companies constituted a relatively profitable investment. For example, the sector average in the "happy twenties" fluctuated at around 6% among those who obtained positive results and some even gained higher indices. The latter were usually located in the large industrial areas and many of them were versatile and operated in other sectors, such as electricity. This situation enabled them to be more competitive, although not all of them obtained high dividends. There was a decreasing trend in the number of public limited companies in the water sector. On the contrary, the prominence of the municipalities that regained the control of this public service increased. In any event, the presence of the private companies was maintained in the management of the supply of water in certain towns. Some of these companies diversified into other sectors, sometimes related to the distribution and consumption of water (packaging and sale of mineral water, spas, pipes, and accessories), others to unusual activities such as medical care. The *Sociedad General de Aguas de Barcelona*, is, perhaps, the most evident example of this.

Andalusia constituted a model in which private companies coexisted with the direct management of the local governments. In 1950, municipalised water services already existed. The few places where the water supply was still in private hands at that time were not relevant and were mainly small towns. In other cases, in the same town, the municipal service coexisted with the concession to a private company, which almost always ended up being absorbed by the local government. In the period between 1850 and 1950, the water supply sector took shape, based on the abundant legislation arising regarding hygiene and sanitary issues related to cities and population nuclei in general.

An historical analysis enables us to visualise the economic possibilities of the sector, particularly in terms of whether there was a sufficient forecast of the needs of the urban areas in the medium term. Several companies showed their restructuring and expansion capacity and knew how to extend their investments to other sectors, some barely related to the supply of water. In Spain, the actions taken in this respect were particularly manifested in the 1960s and

1970s, when large works were undertaken in order to supply running water to all types of towns, even very small ones. These were the years of the National Water Supply and Sanitation Plan, which opened new perspectives in this field. The advances were uniquely relevant, but the lag and delay of the sector was so considerable that the needs were not fully met. A new phase began in 1990 with the legislation referring to the treatment of wastewater and nature conservation. These actions, which were relatively new in this phase, required new economic and financial perspectives.

In short, several conclusions can be extracted from the evolution experienced by the implementation of water supply companies in Andalusia. First, it has been characterised by a "return trip" process towards the privatisation of the sector. The private companies arrived in the sector. This was followed by a second phase of municipalisation and then in the last few decades there has been a trend towards privatisation again. Second, the presence of foreign capital has been noteworthy, particularly of the French and British companies which were predominant in the sector. Finally, the sector has been highly dynamic, with continuous changes in the ownership of the management and frequent take-over and merger processes.

References

Ariño, G. 1993. *Economía y Estado. Crisis y Reforma del Sector Público*. Madrid: Marcial Pons.

Ariño, G. 1997. *El nuevo servicio público*. Madrid: Marcial Pons.

Castro-Valdivia, M., and J. M. Matés-Barco. 2020. "Los servicios públicos y la inversión extranjera en España (1850–1936): Las empresas de agua y gas." *História Unisinos* 24 (2): 221–39. https://doi.org/10.413/hist.2020.242.05.

Castro-Valdivia, M., J. M. Matés-Barco, and M. Vázquez-Fariñas. 2020. "Capital extranjero y desarrollo urbano en España: la inversión en los servicios públicos durante el primer tercio del siglo XX." In *Las infraestructuras de la construcción de la ciudad moderna. España y México, 1850–1936*, edited by L. E. Otero-Carvajal, 177–93. Madrid: La Catarata.

Cuervo, A. 1986. "La privatización de la empresa pública: la nueva desamortización." *Papeles de Economía Española* 27: 331–40.

Fernández-Paradas, M., and J. M. Matés-Barco. 2022. "Ingenieros y empresarios en las compañías de abastecimiento de agua en España (1840–1930)." *Aportes. Revista de Historia Contemporánea* 108, XXXVII: 75–111.

Heredia-Flores, V. M. 2013. "Municipalización y modernización del servicio de abastecimiento de agua en España: el caso de Málaga (1860–1930)." *Agua y Territorio / Water and Landscape* 1: 103–17. https://doi.org/10.17561/at.v1i1.1038.

Heredia-Flores, V. M. 2018. "El abastecimiento de agua en Málaga (1860–1930). De negocio privado a servicio público." In *Agua y Servicios Públicos en España y México*, edited by J. M. Matés-Barco and J. J. P. Rojas-Ramírez, 121–54. Jaen: UJA Editorial.

Martín Mesa, A., F. Alcalá Olid, M. I. Ortega Díaz, J. R. Lanzas Molina, J. J. Duro Cobo, and I. Herrador Linde. 2006a. "Las cuentas del agua en Andalucía: sectores de actividad que emplean el agua como factor esencial en sus procesos productivos." *IX Encuentro de Economía Aplicada*. Jaen.

Martín Mesa, A., F. J. Sáez Fernández, F. Alcalá Olid, F. González Gómez, M. I. Ortega Díaz, J. R. Lanzas Molina, M. A. Sánchez Domínguez, M. A. Ortega Almón, J. J. Duro Cobo, and I. Herrador Lindes. 2006b. "Las cuentas del agua en Andalucía.". *IX Encuentro de Economía Aplicada*. Jaen.

Martínez López-Muñiz, J. L. 1997. "La nueva regulación local en España." In *El Nuevo servicio público*, edited by G. Ariño, J. M. De la Cuétara, and J. L. Martínez López-Muñiz, 185–270. Madrid: Marcial Pons.

Matés-Barco, J. M. 1997. "Las sociedades anónimas de abastecimiento de agua en Andalucía: una primera aproximación." *Boletín del Instituto de Estudios Giennenses* 167: 103–27.

Matés-Barco, J. M. 2002. "Strategies of Foreign Firms in the Sector of Water Supply in Spain (1850–1990)." In *Transnational Companies, 19th-20th Centuries*, edited by H. Bonin, 301–16. Paris: Plage.

Matés-Barco, J. M. 2004. "The Development of Water Supplies in Spain: 19th and 20th Centuries." In *Urban Growth on Two Continents in the 19th and 20th Centuries: Technology, Networks, Finance and Public Regulation*, edited by A. Giuntini, P. Hertner, and G. Núñez, 165–78. Granada: Comares.

Matés-Barco, J. M. 2006. "Las empresas de abastecimiento de agua en Andalucía (1840–2000)." In *El debate del agua en Jalisco y Andalucía*, edited by J. R. Navarro and J. Regalado, 45–93. Sevilla: Junta de Andalucía.

Matés-Barco, J. M. 2018a. "La distribution de l'eau dans les villes d'Espagne (1840–1936): Le rôle des compagnies privées." *Histoire, Économie & Société* 3: 14–29. https://doi.org/10.3917/hes.183.0014.

Matés- Barco, J. M. 2018b. "De la regulación a la privatización y viceversa: la gestión del agua en España y Gran Bretaña." In *Agua y Servicios Públicos en España y México*, edited by J. M. Matés-Barco and J. J. P. Rojas-Ramírez, 29–68. Jaen: UJA Editorial.

Matés- Barco, J. M. 2018c. "Ecología y servicios públicos: nuevas perspectivas y cambio de paradigma." In *Agua y Ecología Política en España y México*, edited by A. Torres-Rodríguez and E. Moral-Pajares, 41–85. Jaen: UJA Editorial.

Matés-Barco, J. M. 2019a. "Estrategia empresarial y control del mercado: la gestión del abastecimiento de agua potable." In *Empresas y empresarios en España. De mercaderes a industriales*, edited by J. M. Matés-Barco, 223–40. Madrid: Pirámide.

Matés-Barco, J. M. 2019b. "El abastecimiento de agua a Barcelona (1850–1939): Origen y desarrollo de las compañías privadas." *Historia Contemporánea* 59 (1): 165–94. https://doi.org/10.1387/hc.19434.

Matés-Barco, J. M. 2021a. "Small, Medium and Large Companies in the Supply of Water in Spain (1840–1940)". In *Entrepreneurship in Spain. A History*, edited by J. M. Matés-Barco and L. Caruana de las Cagigas, 82–98. New York-London: Routledge. https://doi.org/10.4324/9781003126973-7.

Matés-Barco, J. M. 2021b. "Public Services in Spain (1840–1936): The Role of Water Supply Companies." In *Companies and Entrepreneurs in the History of Spain. Centuries Long Evolution in Business since the 15th Century*, edited by M. Vázquez-Fariñas, P. P. Ortúñez Goicolea, and M. Castro-Valdivia, 135–59. Cham (Switzerland): Palgrave MacMillan. https://doi.org/10.1007/978-3-030-61318-1_9.

Núñez Romero-Balmas, G. 2005. "Water Management in Spain: Ecological and Economic Foundations of Institutional Change." In *The Institutional Arrangements for Water Management in the 19th and 20th Centuries*, edited by J. C. N. Raadschelders, 95–118. Amsterdam: Ios Press.

Núñez Romero-Balmas, G. 2018a. "Las aguas como proceso histórico." In *Agua y Servicios Públicos en España y México*, edited by J. M. Matés-Barco and J. J. P. Rojas-Ramírez, 121–54. Jaen: UJA Editorial.

Núñez Romero-Balmas, G. 2018b. "La gestión de las aguas en España; fundamentos ecológicos, económicos y políticos del cambio institucional." In *Agua y Ecología Política en España y México*, edited by A. Torres-Rodríguez and E. Moral-Pajares, 87–118. Jaen: UJA Editorial.

Ponzellini, A. M., and T. Treu. 1994. *Servizi pubblici, privatizzazione e relazioni industriale in Europa*. Milano: FrancoAngeli.

Roversi Monaco, M. G. 1996. *Dominio pubblico e impresa nella gestione dei servizi idrici. L'esperienza in Gran Bretagna e le prospettive in Italia*. Padova: Cedam.

CHAPTER 2

An Incessant Thirst: The Water Market in a Dried-Up Southern European Country, Spain, 1985–2022

Jesús Mirás Araujo and Nuria Rodríguez Martín

Abstract

Spain is a hydrologically unbalanced country, in which water policies have historically focused on increasing supply, through the construction of infrastructures to cope with the irregularity of resources, despite being economically and ecologically unsustainable. The medium-term evolution of water management has been conditioned by the change in the political regime, and by the (recent) changes in the culture of water. The aim of the chapter is to analyse the transformations of regulation that have occurred since the rise of democracy in Spain (in particular since the passing of the 1985 Water Act), the map of water uses, the levels and structure of consumption, and the main agents operating in a sector that in recent decades, despite the public nature of the utility, has undergone an increasing privatisation.

Keywords

consumption – regulation – Spain – water – twenty-first century

1 Introduction

The importance of water in the twenty-first century is undeniable, so much so, that terms such as "blue gold" are becoming widely used to refer to it. Despite the unfortunate economic-financial connotation of the term "resource", which implies the use of water as a profitable asset involving transnational capital, numerous international campaigns have fostered its consideration as a human right (Barlow and Clarke 2002), culminating in its recognition as such by the General Assembly of the United Nations in 2010.

Notwithstanding its geographical heterogeneity, from a territorial point of view, Spain's overall endowment of water resources is unbalanced. The water

 | DOI:10.1163/9789004679634_003

supply is not *strictu sensu* insufficient, but it is irregularly distributed, and is characterised by the contrast between a predominantly humid northern half of the country and a dry south. This lack of balance is a consequence of its geographical position and the characteristics of the environment. Almost three quarters of the country's surface water resources are located north of a hypothetical dividing line between the basins of two of the main rivers of the Iberian Peninsula, the Duero, and the Ebro. Hence, Spain constantly experiences recurring droughts and frequent situations of aridity and water stress, especially in the southern half and, more seriously, in the southeast. Proof of this is that Spain had the third highest water exploitation index in the European Union at the end of the last decade, only behind Malta and Cyprus (Suárez 2020).

Historically, water policies have been overwhelmingly oriented towards increasing supply. The instrument that has been most used is the building of infrastructures to address the irregularity of the resource, to the detriment of more modern and efficient alternatives (although economically and environmentally costly, such as desalination, recycling, etc.), despite the fact that the former option is economically and ecologically unsustainable in the medium and long term. For this reason, until the end of the twentieth century, the predominant solutions were mostly focused on transfers between river basins.

However, the evolution of water management in Spain in recent decades has been largely determined by the change in the political regime, with the return to democracy, and by the changes derived from the "new water culture". The Spanish constitution enacted in 1978, which underlines the obligation of the public authorities to ensure the rational use of all natural resources and protect and restore the environment, marked a milestone. On the one hand, because it gave rise to a new focus on the use of certain resources and on the environment by recognising its importance in the development and quality of life of citizens. On the other hand, because, following the constitution, a new legal framework for water began to be built, related to the changes that later took place in the territorial organisation of the state.

The aim of this chapter is to analyse the transformations that have occurred in the regulation and management of water in Spain since the rise of democracy (particularly since the passing of the Water Act of 1985), the water-use map and the main operators that work in a sector which, despite the public nature of the utility, has undergone an increasing privatisation process, although in recent years this trend seems to have moderated and even changed direction.

2 The Hydrographic Constraints of a Southern European Country

Spain has an unbalanced hydrological system (Del Moral 2009). Due to its location and the characteristics of the environment, the climatic and rainfall conditions are heterogeneous, typical of the Mediterranean climate prevailing in a large part of the territory. In addition, aridity, together with droughts and floods, are recurrent phenomena that are more characteristic of countries of northern Africa than of Europe. In fact, Spain is the most semi-arid country in the European Union (López-Gunn et al. 2012, 83). Although annual rainfall is not too far below the EU average, the problem lies in the irregularity in its geographical distribution. This means that the capacity of regular streams per surface unit is just 4% of that of France, which limits the whole supply (Naredo 2006, 14).

These features are complemented and aggravated by the absence of large rivers in the Iberian Peninsula (which also have irregular flow regimes), or other masses of fresh water,[1] such as lakes or tarns (Giner 1994, 441–442), which has generated "recurrent hydrous tensions", that become more acute in periods of drought (Llamas 1999, 283). The result is the considerable imbalance in the distribution of water in terms of both territory and time. According to Fanlo (2002, 45), Spain is only able to obtain 8% of the resources that exist naturally in the hydrographic network as a whole, that is, without artificially altering them, compared to 40% of other European countries.

As a result, most of the territory and the river basins have a mature water economy, similar to many other regions of the planet. This process began in the 1980s (Gómez and Calatrava 2016, 15), manifesting in an increasing insufficiency of the resource, which is correlated with its scarcity. However, this problem is mainly of a social nature, caused by the rising demand for water from human activities (Aguilera 1997). For more than a decade, this demand and a limited long-term supply have caused the progressive "closure" of most of the river basins in the south and east of the Peninsula (Palomo and Gómez 2013), a situation that is characterised by the lack of sufficient margin to satisfy new water demands without reducing the existing ones. This is relevant for understanding the transformations that have taken place in the Spanish water policy and in the global strategic reorientation of water management companies, particularly multinationals (Babiano 2015).

---

1 In 2007, the Spanish Ministry of the Environment identified important groundwater bodies, which were estimated to cover some 350,000 km^2 (Manzano and Lambán 2012, 35). Underground reserves cover approximately 150,000 to 300,000 hm^3 (Villarroya et al. 2014, 222).

3 Changes in Water Culture

The exploitation of water as a natural resource in Spain experienced a quantitative and qualitative leap in the nineteenth century, as the State expanded its competencies over the resource and the infrastructures by passing the water acts of 1866 and 1879, which enabled a more intensive exploitation. On an international level, in the last half century, steps have been taken to achieve more sustainable water development. This began with the International Covenant on Economic, Social and Cultural Rights, a multilateral treaty voted in the General Assembly of the United Nations on 16 December 1966, which, however, did not become effective until 3 January 1976. Although it does not expressly mention water in the articles, the rights for which it advocates constitute the pillar that has inspired the holding of many international conferences, and the elaboration of declarations and legal proposals that have marked the fight for the protection of water and the guarantee of its access in appropriate and equitable conditions worldwide (Embid 2007).

Subsequently, in the Brundtland Report, presented at the United Nations in 1987, the principle of sustainable development was mentioned for the first time. This was the mainstay on which the European Sustainable Development Strategy (passed in 2001) was based years later, and which has undergone further revisions (Rosado 2010, 49). Precisely, the EU has a long legal tradition, rooted in the declaration of principles that represented the European Water Charter drawn up by the Council of Europe in Strasbourg in 1968 (Berga 2010, 8). The subsequent European directives were inspired by the deliberations of the International Conference on Water and the Environment (IGWE). The results were the conclusions expressed in the Dublin Declaration, a technical meeting prior to the United Nations Conference on the Environment and Development (UNCED) that took place in Rio de Janeiro in 1992. Likewise, several events held since the 1990s, such as the European Conference on Sustainable Cities (the ninth was hosted in Mannheim in 2020), which resulted in the Charter of European Cities and Towns Towards Sustainability or the Aalborg Charter (1994); or the Conference of European Sustainable Cities held in Hannover in 2000, from which the Hannover Declaration emerged, all proposed initiatives aimed at sustainability in European cities.

On the other hand, since 1972, several Environment Action Programmes (EAPs) have been approved. The first came into effect in 1973 and the subsequent programmes have continued to be implemented approximately every five to six years to the present day with the 8th EAP currently in force (2022–2030). However, until recently, EU water law, which has been implemented through Directives (establishing regulatory mandates that leave the final decision on how to comply with them to the states), has been unsystematic and unplanned (Embid 2006, 57–58), leading to its asymmetric incorporation into national laws. However, this has not prevented water law from undergoing a global transformation, given the awareness of its importance in protecting the environment and in recent decades it has become a tool for the protection and conservation of the resource (Embid 2011).

Under EU policy, the Spanish Strategy for Sustainable Development was designed in 2007, with the goal of saving and efficiency in the use of resources. It was proposed in parallel with another instrument, the National Plan for Adaptation to Climate Change (PNACC), a series of measures related to water use and management policy (Rosado 2010, 51). However, in recent years, expectations have only improved slightly. International commitment to climate change, sustainable development strategies and circular economy approaches are changing the planning, use and management of water (Maestu and Del Villar 2019, 339).

In Spain, the "New Water Culture" emerged in the mid-1990s, although its completion is proving difficult. Progress has been slow and discontinuous, generating a certain degree of frustration due to the inability of the central government to change the structure of the sector (Aguilera and Martínez 2007, 294). The defenders of this new culture believe that Spanish hydrological planning has not advanced enough as a result of being "kidnapped" by corporate capitalism, that is, by construction and electricity companies, which have imposed formulas and solutions to water problems (Aguilera 2014).

The new focus of water planning, which is more comprehensive and demand-oriented, requires new tools and instruments. Gómez and Calatrava (2016, 17) allude to regulatory and economic instruments. The former includes the modification of water use rights. However, it has been acknowledged that they lack effectiveness, which is why more recently economic instruments are becoming more widely used. They allow users to alter the demand for water by means of economic incentives that enable certain uses to be reduced. These instruments include the implementation of new legal figures and institutions that reorganise formal water markets. These markets were introduced into the Spanish legal system with the Act 46/1999, amending Water Act 29/1985. Among the latter, the contracts for the transfer of water rights (inter-basin or intra-basin) or the exchange centres for use rights (known as water banks) are predominant, even though they have had relatively little activity (Maestu and Gómez 2010, 288).

4 Transformations in Regulation since the Coming of Democracy

The recent evolution of water management has been conditioned by the change in Spain's political regime. The Franco dictatorship was "continuist" in terms of water and was marked by the persistence of the regenerationist projects implemented during the first third of the twentieth century and by the strong centralisation of management through the Hydrographic Confederations (river basin confederations), which were created in 1926.

After the 1978 constitution, a new approach to address the use of certain resources and the environment was introduced and a new legal framework began to be built. Although the constitution does not regulate water as a natural resource or recognise its use as an autonomous and independent right, article 45 recognises the right of citizens to enjoy an adequate environment as well as their obligation to conserve it. Article 132 refers to state public domain assets, establishing the distribution of water competencies between the State and the nascent Autonomous Regions. One of the first measures was the remodelling of the Hydrographic Confederations, through Royal Decree 2419/1979, of 14 September, which determined the composition and functions of the governing bodies of those river basin confederations (Melgarejo 2000, 91).

Following the period of the first democratic centre governments, the first socialist government (1982) initiated the change in water regulation. The first action taken was the Water Act (29/1985), which assigned the water competencies to the State. It established the public water domain, which is made up of surface continental waters and renewable groundwater. The Act also ratified unitary management and the existence of a single body in each basin, to which all functions were assigned (Matés-Barco 2016, 21), establishing water planning as the instrument for basin management. Likewise, it began to propose the integration of water management with land use planning and discussed (but not applied) concepts such as water economics, decentralisation, efficiency and users' participation and a proper understanding of the water cycles (Ortega 2012, 89).

The 1985 act underwent subsequent revisions, although its essential principles were not substantially altered. In 1992, the National Water Council was created, constituting the highest advisory body on water issues and an instrument for expressing the opinion of the institutional actors recognised by the law: the Administration of the State and the Autonomous Regions, the Basin Organisations, the local entities, and the professional, economic and ecological organisations with a national scope related to water uses (De Cisneros 2009, 188–189).

Nevertheless, the first major change dates back to 1993, when the Draft Bill of a new National Hydrological Plan was approved, the key piece of which was the Integrated System for National Hydraulic Balance (SIEHNA). The plan was based on principles of inter-territorial solidarity and compensation, with which to alleviate the unequal distribution of water resources, in line with the regenerationist ideal upheld by the Spanish political leaders during the twentieth century. The SIEHNA sought a substantial water restructuring, correcting the hydrological imbalances between territories, but maintained a "continuist" approach regarding water management planning, still focused on supply and based on the construction of reservoirs and hydroelectric dams (Mateu, 2008). To do this, it was to construct 150 reservoirs and redistribute 3,768 million m^3, through transfers between basins (Gómez and Del Moral 1995), repeating the Francoist policy of building reservoirs. The result was an overdevelopment in the construction of surface water reservoirs: at the end of the last century, Spain was the country with the highest proportion of relative area covered by reservoirs in the whole world (Naredo 1999). However, the plan did not succeed due to political and social reasons. The slowdown in economic growth after the 1993 crisis was a major, albeit transitory, obstacle. The main cause was the strong social opposition to the project (Aguilera 1997, 235).

In 1999, a new Water Act was passed (Act 46/1999, of 13 December), which introduced some changes in the approaches related to water management. For instance, the consideration of environmental flows as a prior restriction to other uses and the introduction in the legislation of the concept of good ecological status of water (Hernández et al. 2011, 3). It also addressed environmental issues, such as waste regulation, wastewater recycling, and desalination of sea water. However, the most important novelty was the introduction of a new economic tool, the contracts for the transfer (assignment) of water use rights, which would be considered as temporary and constituted the first step towards the creation of an incipient water market on a national scale (Martín 2000). However, a water market, as such, was not created, but rather a market for water use rights. The Act assigned a central role to the Basin Organisations in these markets since the water transfer contracts are explicitly or implicitly approved by these organisations. However, the general framework of water policy continued to focus on increasing the water supply to meet the rising demands (Cabrera et al. 2003, 248–50). The government sought to coordinate this objective with the conservation of the resource, but this is a duality that is difficult to address, and that continues to characterise (and hinder) water policy in Spain today (Hernández et al. 2011, 3).

In 2001, through Royal Legislative Decree 1/2001, of 20 July, the Recast Text of the Water Act was passed. Spain's entry into the European Community (1986) forced it to adapt to the legislative changes derived from the progressive construction of a European water policy, which was materialised in the development of the EU Water Framework Directive, WFD (Directive 2000/60/EC). Before its entry into force, there had already been a European regulation on water, namely the Directive 91/271/EEC of the Council of Europe, of 21 May 1991, concerning urban waste-water treatment. However, the former was later completed with other legislative provisions regulating specific aspects of water use.

The WFD represented a qualitative leap in environmental issues with respect to previous legislation (Pérez 2015, 32), and came about the recognition of the need to formalise the application of economics in water management, through the design of policies that apply economic principles, methodologies, tools and instruments (Pulido, Cabrera, and Garrido 2014, 106). Although its real impact has probably been overvalued, since it has generated "a collective illusion", according to Cabezas (2012), it is probably the most ambitious and complex of all European environmental regulations.

The Directive was built upon several core ideas: environmental sustainability, economic rationality and the principle of efficiency, transparency, and social participation (Hernández 2013, 42). These ideas hold several fundamental principles: 1. Water is not a commercial good like any other, but rather a heritage that must be protected and defended; 2. Water is a scarce natural resource; 3. Water is an economic good that is likely to be exploited; 4. Water is part of the territorial environment. Based on the former principle, the WFD recognises the natural geographic framework of the hydrogeological cycle of inland waters and establishes the hydrographic basin as the territorial framework for water management (Pérez 2015, 32). Its implementation brought about the design of iterative six-year planning processes, with the establishment of a precise calendar for its execution (La Roca 2018, 3).

The WFD has been inadequately transposed into the Spanish national legislation, as the new priorities have overlapped with the traditional objectives of the national water policy, which had traditionally been focused on supply (Hernández 2013, 40). The legal adaptation was made through the Act 62/2003, on Fiscal, Administrative and Social Order Measures, by which the text of the Water Act was partially modified.

However, in parallel, Act 10/2001, of 5 July on the National Hydrological Plan (NHP), embodied an approach that radically opposed that of the WFD, since it contemplated a model of authoritarian water management that recovered the long-established idea of water transfers from the "surplus" basins to the deficit ones (Aguilera 2003). Its gestation was long, starting in 1996, with the elaboration of several government drafts, and a White Paper on Water in 1999. But it did not have social or political consensus, as evidenced by the long parliamentary debate prior to its sanction (Bosque 2008, 475–476). The Plan was designed to constitute the main instrument of the national water policy (Melgarejo and López 2015, 138).

However, the application of the plan was blocked due to the widespread social rejection in the regions that were affected by the water transfers and by the political change that took place in Spain after the general elections of 2004. Finally, Royal Decree Act 2/2004, of 18 June, repealed the flagship project of the NHP, the transfer of the Ebro River, of 1,050 hm^3, which sought to resolve the water problem of the Mediterranean arch (which had an estimated deficit of 2,015 hm^3).

The next legislative action was Act 11/2005, which modified Act 10/2001 of the National Hydrological Plan. The new strategy culminated in the approval, that year, of the A.G.U.A. Programme, "*Actuaciones para la Gestión y la Utilización del Agua*" (Actions for the Management and Use of Water), which was, in theory, more focused on recycling and desalination, with a time horizon for 2004–2008. The instrument to achieve the goal was the building of new desalination plants, with would contribute an estimated 1,063 hm^3 to the Mediterranean basins (Mateu 2008). Although the new plan seemed to introduce a new vision and a new approach to water policies, in practice the programme implied significant investments in new catchment facilities, with less emphasis on demand policies and on sustainable development. In addition, the decree maintained the oversized estimates of water needs and demands made by the previous NHP (Aguilera 2005, 268). Overall, these Acts did not substantially alter the basic water principles established by the 1985 Act.

From an institutional viewpoint, water management in Spain is a public responsibility, although it is highly fragmented. The distribution of competencies is complex, typical of a decentralised State, where numerous agencies are involved in water administration, thus generating overlapping problems: Government, National Water Council, Ministry of the Environment, River Basin Organisations (Hydrographic Confederations), Water Administrations of the Autonomous Regions, city councils and irrigation communities. The reforms of the statutes of several Autonomous Regions during the second half of the 2000s extended the competences of the Spanish regions, except in those issues over which the State has exclusive competence (Aura 2007, 452).

The global management of water in the basins is carried out by the Basin Organisations. Currently, Spain has twenty-five river basin demarcations. The state manages twelve inter-community basins, through the Hydrographic Confederations. Its territorial scope covers several Autonomous Regions, which is why these are basins of state competence. Conversely, the basins that are only in the territory of one Autonomous Region constitute the so-called intra-community basins (thirteen), in which the regional governments have assumed responsibility for water management. In addition, there is the Eastern Cantabrian river basin demarcation, with shared jurisdiction (state and regional) (Navarro and Estruch 2022, 321). Likewise, there are also demarcations corresponding to the basins that are shared with other countries (Portugal and Morocco).

According to the calendar established by the WFD, in recent years, water plans for river basin demarcations have been implemented in Spain during two water planning cycles (2009–2015 and 2015–2021). The planning cycles are prepared for six-year periods, based on a diagnosis of the situation, the determination of objectives, the development of programmes of measures and the establishment of a monitoring and control programme (Hernández 2013, 44). However, the first planning cycle accumulated considerable delays. These delays also had consequences for the second planning cycle, since there was not enough time to implement the programmes of measures or review their degree of compliance (Castro 2015, 134). Finally, of the twenty-five river basin demarcations, eighteen managed to approve their water plan for the second planning cycle, although their assessment is questionable from the point of view of their effectiveness (González 2018).

Driven by Royal Decree 907/2007, by which the Hydrological Planning Regulation Act was approved in January 2020, the preparation of the "Provisional Schemes of Important Topics of the water planning process" (PSIT) began for the cycle 2021–2027 (Caro and Escartín 2021, 666). At present, the projects of water plans of the twelve inter-community river basin demarcations have been sent for processing to the Ministry for Ecological Transition and Demographic Challenge, while those of the intra-community demarcations are in the consultation period.

Finally, in Spain there are also the so-called State Water Companies (nine at present), which were constituted under the Act 13/1996, of 30 December on Fiscal, Administrative and Social Order Measures. These firms are registered as private companies, with a structure similar to a public limited company, although they are one hundred per cent owned by state capital and must account for their management before the Court of Auditors prosecutors' office. The legal mechanism by which they operate is through a concession contract for the building and exploitation of water works. Their peak period was 1997–2004, during the conservative government.

5 Map of Water Uses

In order to understand the water uses, it is necessary to briefly examine the structure of water management. Two major phases can be distinguished in the global water cycle. The first, called headrace or "upstream" distribution, corresponds to the activities of "abstraction", the storage in reservoirs and the transport of water to the municipalities. Its management and regulation correspond to the river basin confederations, which manage the riverbeds, plan investments, and operate the facilities. In this phase, agricultural consumption and that of large industries or power plants are managed. The second phase, called "*agua en baja*" (retail water supply) or urban water cycle, begins with the transfer of water to the municipalities, where it is made drinkable, distributed to residential users, collected through the sewage network and treated in wastewater treatment plants for its return to the natural channels. The regulation and the rates paid by the users in this phase correspond to the municipalities and the Autonomous Regions, which are the owners of the infrastructures (Price Waterhouse Coopers 2018, 7).

However, at present, there is no single regulation at the national level. Rather, the regulatory system is dispersed among the more than 8,000 municipalities that are responsible for providing the water supply, which results in a complex regulatory framework without a clearly dominant authority (Estruch 2021, 281). In addition, for towns with a population of less than 20,000 inhabitants, the Provincial Governments coordinate the water supply to the users and the draining and treatment of wastewater, either through direct provision or by shared management methods (consortiums, business associations, etc.) (CNMC 2020, 17). In fact, there is an increasing tendency to provide this utility at a supra-municipal level, which has been legitimising the intervention of other Administrations, especially the regional ones (Calvo 2010, 303–304).

As a whole, the most extensive network is the water supply system, which has a length of 248,245 km, followed by the drainage network (189,203 km), and the headrace network (23,789 km). However, the whole network will require a major investment in the future for its renovation, which has been identified and quantified by the Ministry for Ecological Transition at a total of 45,192 million euros until 2033 (Pérez and Ballesteros 2020, 155 –157).

Spain's water use map differs from that of most developed nations, although it is similar to that of other countries in Southern Europe. There is a high concentration of demand in agricultural uses. According to the latest survey on water supply and drainage, in 2018, 15,494 hm^3 were supplied to farms. On the other hand, urban uses are usually divided into domestic uses, which take place in households, and non-domestic uses, which include industry, commercial activities and public uses. However, urban uses constitute a very heterogeneous ensemble, which reflects the historical conditions of development of each city and its water infrastructure (Martínez 2014, 5). A total of 4,236 hm^3 were allocated to public urban supply networks in 2018. Of these, 2,271 hm^3 were recorded as consumption by households, 629 hm^3 by economic sectors and 288 hm^3 corresponded to municipal consumption (3,732 hm^3 in total). These figures show a significant containment of consumption in these three segments with respect to the data from a decade earlier (2,540, 833 and

359 hm³, respectively, in 2008). The average household water consumption stood at 133 litres per inhabitant per day, a figure considerably lower than the 154 litres recorded in 2008. According to the origin of the water collected by the private companies and the public entities, excluding purchases to third parties, 64.3% was distributed as surface water, 30.9% as groundwater and the remaining 4.8% as other types of water (desalinated seawater or brackish water).[2] Annually, approximately one fifth of the aquifers are allocated to urban uses (De Stefano *et al.*, 2015), although in towns with less than 20,000 inhabitants the percentage is around 70% (Garrido and Llamas 2009). Therefore, the primary source of water in Spain is surface water, although groundwater and desalination are progressively gaining share.

6 Main Operators in Water Supply and Changes in the Management

By the late 1970s, a wave of liberalisation, deregulation and privatisation of public companies began in the Anglo-Saxon economies, as a result of which many industrialised countries changed their legislation to facilitate the outsourcing of public utilities (Ruiz, González, and Picazo 2015). Consequently, since the 1980s there has been a progressive neo-liberalisation of natural resources, which has led to a global commodification and privatisation of water (Swyngedouw 2007), with the additional effect of an increasing process of business concentration around the so-called multi-service multinationals.

The 1990s paved the way for new strategies for the private provision of the service and for new corporate and investment strategies by the large international water operators, which were based on an intensive process of capital accumulation. This facilitated the emergence, during the turn of the century, of new financial players in the water markets on a global scale. This "financialisation" process has become consolidated since the first decade of the twenty-first century (Loftus, March, and Purcell 2019).

The privatisation of the water supply can be explained, from a theoretical viewpoint, by the alleged advantages in terms of efficiency, costs, tariffs and even quality of the service with respect to public management, although the results do not seem to have met expectations. For this reason, experts, academics, managers (led, especially, by certain political parties) and civil society, have promoted a movement to remunicipalise water. This change began in some

2 Instituto Nacional de Estadística. *Estadística sobre el Suministro y Saneamiento del Agua. Año 2018*, 2020.

large European cities (Paris, Berlin...) in the mid-2010s (Ruiz and García 2017, 2480–1).

In most of the EU countries, the management of urban water supply, sewerage and wastewater treatment services is carried out mainly by the municipalities (Babiano 2015, 134). In Spain, the water supply for irrigation is carried out by individual users, or by irrigator communities or irrigation collectives. Conversely, Act 7/1985 on Local Regime Bases establishes that urban water supply, sewerage and wastewater treatment services are the responsibility of municipalities, which will provide them as a monopoly or with exclusive rights. These are public utilities that must obligatorily be provided, although the decision on the legal entity that supplies the service is in the hands of the local councils. The formulas are diverse and range from the direct provision of the service by the municipality to the transfer of its management to public or private companies, or to other forms of private-public collaboration. When urban water supplies are provided directly by the public sector, the most common formulas are municipal services, associations, consortiums of municipalities or public companies (Arbués and García 2020, 7).

One of the legacies of the 1970s crisis in Spain was a delicate budgetary situation, which harmed the municipalities by downsizing their volume of financial resources, making water privatisation an attractive option (Ruiz 2013, 36). As a result, since approximately 1985, a gradual transfer of management to private companies has been observed in many municipalities, a trend that intensified during the following decade (Ruiz, García, and González 2010).

The process could take two forms. First, by the transfer to private companies under a concession regime. In this system, the local government retains ownership of the service but entrusts its management to a person or legal entity for a contractually agreed period. Second, by a partial privatisation, through a joint venture. The most frequent is the first type (García, González, and Picazo 2013, 60). These concessions are awarded through public tender and for a limited time, and the municipality charges the company a concession fee. In the case of water supply companies, if the contracts involve the building of infrastructures and the operation of the service, their duration must not exceed fifty years, while those that only involve operating the service have a restriction of twenty-five years (García-Valiñas 2019, 42).

The result of the process is that of all the developed economies Spain is among those with the greatest prominence of private companies in the water industry. In 1996, 63% of the Spanish population was supplied by some form of public management, and 37% by private management (Babiano 2015, 135). In 2002, according to the *Spanish Association of Water Supply and Sanitation* (AEAS), 40% of the population was served by a private concessionary company, this system being predominant in cities with more than 100,000 inhabitants (Del Villar and Maestu 2006, 45). On the other hand, according to the eleventh survey on the supply of drinking water and sanitation in Spain (AEAS, 2010), in 2009, 43% of the population was supplied by public companies, 33% by private companies, 13% by joint ventures, and 7% directly by the municipalities. In towns with less than 100,000 inhabitants, private management prevailed, while in those with larger populations, public, private and mixed regimes coexisted. The subsequent economic crisis accentuated the trend towards privatisation, raising the participation of the private sector to 56% in 2016 (AEAS/AGA, 2016). In 2020, private companies (33%) and joint ventures (22%) supplied 55% of the Spanish population, while 35% were supplied by public companies, with the remaining 10% being supplied by municipal services (Albalate et al. 2022; AEAS/AGA 2020).

According to the Central Directory of Companies of the National Institute of Statistics,[3] the number of firms engaged in water collection, distribution and sewage disposal was 3,241 in 2019. An evidently oligopolistic structure can be observed, with two large groups dominating the sector, *Aguas de Barcelona* (*Agbar*) and *Aqualia*. However, at the same time, the sector is also highly fragmented due to the large number of operators.

In 2018, the corporation with the most significant presence in the sector was *Fomento de Construcciones y Contratas* (*FCC*), which ranked 122nd among the 5,000 largest companies in the country, with revenues of 1,247.46 million euros. *FCC* is an international group of urban utilities, with a presence in more than twenty-five countries, an annual turnover in 2021 of 6,659.3 million euros and a workforce of more than 59,000 employees. It includes, among many other companies, the fourth in the ranking of the sector in Spain, *FCC Aqualia* (the water management company owned by the *FCC* group, 51%, and by the Australian ethical fund IFM Investors, 49%) in position 239, *FCC Medio Ambiente*, *FCC Ámbito* and many more. The second individual water company in the country, in position 193, was *Canal de Isabel II* (the main public company in the sector) and *Urbaser*, owned since June 2021 by the investment capital firm *Platinum Equito* (position 223).

The *Agbar* group, with its headquarters in Barcelona, belongs to the French group *Suez Environnement*, which operates under different business names in several Spanish regions, controlling 54% of the sector, through its branch *Aquagest*.[4] In 2021, it generated an operating income of 3,229 million euros, an

3 Instituto Nacional de Estadística. *Directorio Central de Empresas*.

4 Recently, the France-based multinational environmental services firm, *Veolia*, acquired Suez Spain, the Spanish subsidiary company of Suez Environnement, owner of Agbar. See "Veolia

amount that far exceeded the sum of the sales of the rest of the Spanish business units in this sector. It is, with FCC, the only operator dedicated to global water management that forms part of the main business conglomerates in the country. It manages the water supply in more than 1,400 municipalities, with a population of over 25.5 million people, by controlling some of the most important water operators in Spain (*Sorea, Viaqua, Hidraqua*, etc.). In 2020, these two operators provided the service to two out of every three Spanish municipalities that use private water companies, which is the equivalent of providing the service to three out of every four inhabitants. Meanwhile, the French company *Bouygues-Saur* controls another 13%, with its stake in *Gestagua*. And, finally, ACS, with its subsidiary *Urbaser*, has a share of almost 7%.

By size of the workforce, the fragmentation was actually considerable: 573 companies (17.7%) did not have employees, 1,674 companies (51.7%) had 1–2 employees, and 442 (13.6%) between 3–5 employees. In other words, 83.0% of the companies had five employees or less. As for the largest companies, there were only five companies that fell within the range of 1,000–4,999 employees (0.2%), and seven companies within the range of 500–999 employees (0.2%). Therefore, only 12 companies (0.4%) had more than 500 workers. As a result, the importance of catchments made by individuals and firms that supply small towns can be appreciated. However, there has also been a process of mergers and/or acquisitions, with an increasing presence of large-scale operators.

7 Conclusions

Water management in Spain has undergone profound changes over the last forty years. From an institutional context marked by municipal management, under the protection of the legislation prevailing during the Franco dictatorship, it has shifted to mostly private management, which fits into the neo-liberal paradigm that has been in force over the last two decades on an international level. As a consequence, a substantial segment of the management is in the hands of two large business conglomerates, which operate on a global scale, following the strategies of multi-service transnational groups.

The rise of democracy in Spain and its subsequent entry into the European Community triggered a transformation in its water legislation and policies. On the one hand, the changes in the culture of water, that are still in progress, together with EU water regulations, have forced a progressive reorientation

cierra la absorción de Agbar en España y alcanzará los 15.000 empleados", *CincoDías*, 30 June 2022. https://cincodias.elpais.com/cincodias/2022/06/30/companias/1656586100_446596.html

of water management. An approach that had historically been based on supply and increasing the volume of water supplied has gradually shifted towards policies that are more focused on demand control, although there is still a long way to go, since progress has been desperately slow. In a country that, within the European context, is at serious risk of suffering severe water crises in the future, it is vitally important to properly manage water uses and move decisively towards sustainable policies that preserve the environment in general and water in particular.

These goals have not always been met in the period studied, due to the contradictions between Spanish and EU regulations and the complicated implementation of the European directives, given the variety of vested socioeconomic interests and agents involved. One of the main distortions arises from the high share of water consumption corresponding to agricultural uses, which account for a low proportion of GDP. The new approaches related to the circular economy will be enormously important so as to guarantee the future of the resource in our country.

References

Aguilera, F. 1997. "Economía del agua: Reflexiones ante un nuevo contexto." In *La gestión del agua de riego*, edited by J. López and J. M. Naredo, 235–50. Madrid: Fundación Argentaria-Visor.

Aguilera, F. 2003. "Gestión Autoritaria versus Gestión Democrática del Agua." *Archipiélago: Cuadernos de crítica de la cultura* 57: 34–42.

Aguilera, F. 2005. "El ocaso de la vieja cultura del agua en España: ¿Iniciando la transición hacia una nueva cultura del agua?" *Economistas* 104: 268–72.

Aguilera, F. 2014. "Veinte años (1993–2013) de "política hidrológica en España": el secuestro de la política pública." *Revista de Economía Crítica* 17: 20–43.

Aguilera, F., and J. Martínez. 2007. "La nueva cultura del agua, en espera de otra oportunidad." *Economistas* 111: 292–8.

Albalate, D. et al. 2002. "Legislative reforms and market dynamics in the provision of urban water service by private contract operators in Spain." *Utilities Policy* 74, 101302. https://doi.org/10.1016/j.jup.2021.101302.

Arbués, F., and M. García. 2020. "Water Tariffs in Spain." *Oxford Research Encyclopedias*. https://doi.org/10.1093/acrefore/9780190632366.013.246.

Asociación Española de Abastecimientos de Agua y Saneamiento.2010. *XI Encuesta de Suministro de Agua Potable y Saneamiento en España*. Madrid.

Asociación Española de Abastecimientos de Agua y Saneamiento-Asociación Española de Empresas Gestoras de los Servicios de Agua Urbana.2016. *XIV Estudio Nacional de Suministro de Agua Potable y Saneamiento en España*. Madrid.

Asociación Española de Abastecimientos de Agua y Saneamiento-Asociación Española de Empresas Gestoras de los Servicios de Agua Urbana.2020. *XVI Estudio Nacional de Suministro de Agua Potable y Saneamiento en España 2020*. Madrid.

Aura, A. M. 2007. "El agua en España: un análisis de las recientes reformas estatutarias desde el Derecho internacional." *Anuario Español de Derecho Internacional*, XXIII: 451–85.

Babiano, L. 2015. "Agua urbana: Mercantilización y resistencia social en España." *Agua y Territorio / Water and Landscape* 6: 133–41. https://doi.org/10.17561/at.v0i6.2816.

Barlow, M., and T. Clarke. 2002. *Blue Gold. The Battle against Corporate Theft of the World's Water*. London: Earthscan Publications.

Berga, L. 2010. "La gobernanza del agua en España." *Revista de Obras Públicas* 3507: 7–20.

Bosque, J. 2008. "El agua como recurso escaso y sus problemas en la España actual." *Estudios Geográficos* 265: 453–93. https://doi.org/10.3989/estgeogr.0436.

Cabezas, F. 2012. "The European Water Framework Directive: A Framework?" *International Journal of Water Resources Development* 28 (1): 19–26. https://doi.org/10.1080/07900627.2012.640608.

Cabrera, E. et al. 2003. "Directrices para una política sostenible del agua." *Ingeniería del Agua* 10 (3): 245–57. https://doi.org/10.4995/ia.2003.2584.

Calvo, J. L. 2010. "Abastecimiento de agua potable y saneamiento de las aguas residuales urbanas en España." *Revista Aragonesa de Administración Pública* 36: 295–312.

Caro, I., and V. Escartín. 2021. "Aguas: la nueva gobernanza del agua". In *Observatorio de Políticas Ambientales 2021*, edited by G. García et al., 663–89. Madrid: CIEMAT.

Castro, M. 2015. "Primer ciclo de planificación hidrológica en España en aplicación de la Directiva Marco del Agua." *Agua y Territorio / Water and Landscape* 5: 134–42. https://doi.org/10.17561/at.v0i5.2540.

Comisión Nacional de los Mercados y la Competencia (CNMC). 2020. *E/CNMC/07/19 Estudio sobre los servicios de abastecimiento y saneamiento de agua urbana*.

De Cisneros, J. C. 2009. "La evolución de la política del agua en España en la década de los noventa." PhD diss., University of Castilla-La Mancha.

De Stefano, L. et al. 2015. "Groundwater Use in Spain: An Overview in Light of the EU Water Framework Directive." *International Journal of Water Resources Development* 31 (4): 640–56. https://doi.org/10.1080/07900627.2014.938260.

Del Moral, L. 2009. "Política hidráulica y desequilibrios territoriales en España. Historia y perspectivas." In *Agua, territorio y paisaje: de los instrumentos programados a la planificación aplicada*, edited by L. Sánchez, and M. A. Troitiño, 57–73. Madrid: Asociación Interprofesional de Ordenación del Territorio.

Del Villar, A., and J. Maestu. 2006. "Servicios del agua y recuperación de costes." *Ambienta: La revista del Ministerio de Medio Ambiente* 59: 45–51.

Embid, A. 2006. "El derecho de aguas de la Unión Europea contemplado desde la perspectiva española. Consideración especial de la Directiva Marco Comunitaria

2000/60/CE." In *El agua en el siglo XXI. Gestión y planificación*, edited by J. M. Cuadrat, 57–82. Zaragoza: Diputación de Zaragoza-Institución Fernando el Católico.

Embid, A. 2007. "The Right to Water." *International Journal of Water Resources Development* 23 (2): 267–83. https://doi.org/10.1080/07900620601182968.

Embid, A. 2011. "La transformación ambiental del derecho de aguas: el derecho de aguas del siglo XXI." *Revista de la Escuela Jacobea de Posgrado* 1: 3–19.

Estruch, M. E. 2021. "Diseño de una propuesta para la regulación técnica de los servicios de agua en España." PhD diss., Universitat Politècnica de València.

Fanlo, A. 2002. "La gestión del agua en España: experiencias pasadas, retos futuros." *Revista Electrónica de Derecho de la Universidad de La Rioja* (REDUR) 0: 43–52. https://doi.org/10.18172/redur.3811.

García, M. A., F. González, and A. J. Picazo. 2013. "Is the price of water for residential use related to provider ownership? Empirical evidence from Spain." *Utilities Policy* 24: 59–69. https://doi.org/10.1016/j.jup.2012.07.009.

García-Valiñas, M. A. 2019. "Water Governance in Spain: The Role of Federalism and Public – Private Partnerships." In *Facing the Challenges of Water Governance. Palgrave Studies in Water Governance: Policy and Practice*, edited by S. Porcher, and S. Saussier, 29–55. New York: Palgrave Macmillan. https://doi.org/10.1007/978-3-319-98515-2_2.

Garrido, A., and M. R. Llamas. 2009. *Water Policy in Spain*. London: CRC Press. https://doi.org/10.1201/9780203866023.

Giner, C. 1994. "El problema del agua en España." *Revista de Fomento Social* 49: 437–54. https://doi.org/10.32418/rfs.1994.195.2733.

Gómez, J. A., and J. Calatrava. 2016. "Los mercados de agua y su implementación en España. Una introducción." In *Los mercados de agua en España: presente y perspectivas*, edited by J. A. Gómez, and J. Calatrava, 15–40. Almería: Cajamar.

Gómez, J., and L. Del Moral. 1995. "El Plan Hidrológico Nacional: criterios y directrices." In *La planificación hidráulica en España*, edited by A. Gil, and A. Morales, 331–78. Valencia: Caja de Ahorros del Mediterráneo.

González Rojas, D. 2018. "La gestión de las cuencas hidrográficas en España: avances y carencias del segundo ciclo de planificación." *Agua y Territorio / Water and Landscape* 11: 123–36. https://doi.org/10.17561/at.11.3027.

Hernández, N. 2013. "La planificación hidrológica y la Directiva Marco del Agua en España en el periodo 2000–2012: ¿Hemos avanzado?" In *El río Tajo, lecciones del pasado para un futuro mejor*, edited by B. Larraz, and A. Cano, 39–56. Toledo: Ledoria.

Hernández, N. et al. 2011. "La Planificación Hidrológica y la Directiva Marco del Agua en España: estado de la cuestión." *Papeles Seguridad Hídrica, Agricultura y Naturaleza. Los Nuevos Planes de Demarcación Hidrográfica Según la Directiva Marco del Agua* 2.

La Roca, F. 2018. "La revisión de los planes hidrológicos. ¿Cuáles son los temas importantes?" In *Retos de la Planificación y gestión del agua en España. Informe 2018,* edited by F. La Roca, and J. Martínez, 3–13. Zaragoza: Observatorio de las Políticas del Agua (OPPA).

Llamas, M. R. 1999. "Mitos y realidades sobre la crisis del agua en España." *Arbor* 646: 271–84. https://doi.org/10.3989/arbor.1999.i646.1590.

Loftus, A., H. March, and T. F. Purcel. 2019. "The Political Economy of Water Infrastructure: An Introduction to Financialisation." *Wiley Interdisciplinary Reviews: Water* 6 (6). https://doi.org/10.1002/wat2.1326.

López-Gunn, E. et al. 2012. "Lost in Translation? Water Efficiency in Spanish Agriculture." *Agricultural Water Management* 108: 83–95. https://doi.org/10.1016/j.agwat.2012.01.005.

Maestu, J., and A. Del Villar. 2019. "Financiación y fiscalidad en el contexto de la gobernanza del agua en España." In *Congreso Nacional del Agua Orihuela. Innovación y Sostenibilidad*, edited by J. Melgarejo, 339–56. Alicante: Universidad de Alicante.

Maestu, J., and C. M. Gómez. 2010. "La escasez de agua y el papel de los instrumentos económicos." *Economistas* 123: 286–9.

Manzano, M., and L. J. Lambán. 2012. "Una aproximación a la evaluación de los servicios de las aguas subterráneas al ser humano en España." *Ambienta: La revista del Ministerio de Medio Ambiente* 98: 32–41.

Martín, R. 2000. "El agua como mercancía." *Revista de Administración Pública* 152: 7–27.

Martínez García, A. (ed.). 2014. *La reducción del consumo de agua en España: causas y* tendencias. Madrid: Fundación Aquae.

Matés-Barco, J. M. 2016. "La regulación del suministro de agua en España: siglos XIX y XX." *Revista de Historia Industrial* 61: 15–47.

Mateu, J. F. 2008. "El estado actual de los ambientes del agua en España." In *España y el Mediterráneo. Una reflexión desde la geografía española: aportación española al XXXI Congreso de la Unión Geográfica Internacional*, edited by M. Alario, 35–40. Madrid: Ministerio de Fomento, Instituto Geográfico Nacional.

Melgarejo, J. 2000. "Las confederaciones hidrográficas. Evolución y transformación (1926–2000)." *Alquibla: Revista de investigación del Bajo Segura* 6: 79–95.

Melgarejo, J., and I. López. 2015. "Evolución de la planificación hidrológica en la España democrática, 1978–2014." In *Agua y derecho. Retos para el siglo XXI: Reflexiones y estudios a partir del Water Law*, edited by J. Melgarejo et al., 147–78. Aranzadi Thomson Reuters.

Naredo, J. M. 1999. "El agua y la solidaridad." *Boletín CF+S* 11. http://habitat.aq.upm.es/boletin/n11/ajnar.html.

Naredo, J. M. 2006. "La encrucijada de la gestión del agua en España." In *El agua en el siglo XXI. Gestión y planificación*, edited by J. M. Cuadrat, 9–28. Zaragoza: Diputación de Zaragoza-Institución Fernando el Católico.

Navarro, S., and V. Estruch. 2022. "Implementación de la Directiva Marco del Agua en España: perspectivas futuras. Revisión bibliográfica." *ITEA-Información Técnica Económica Agraria* 118 (2): 318–38.

Ortega, A. 2012. “De aguas, tierras y políticas hidráulicas en la España contemporánea.” *Vínculos de Historia* 1: 73–94.

Palomo, S., and J. A. Gómez. 2013. “El papel de los mercados como instrumento para la reasignación del agua en España.” *Agua y Territorio/ Water and Lanscape* 2: 78–92. https://doi.org/10.17561/at.v1i2.1347.

Pérez Lázaro, R. 2015. *La Nueva Cultura del Agua, el camino hacia una gestión sostenible. Causas e impactos de la crisis global del agua.* Bilbao: Cuadernos de Trabajo/Hegoa Lan-koadernoak no. 68.

Pérez Zabaleta, A., and M. Ballesteros. 2020. “La inversión necesaria en renovar las infraestructuras del ciclo urbano del agua.” *Presupuesto y Gasto Público* 101: 153–66.

Price Waterhouse Coopers. 2018. *La gestión del agua en España. Análisis y retos del ciclo urbano del agua.*

Pulido, M., E. Cabrera, and A. Garrido. 2014. “Economía del agua y gestión de recursos hídricos.” *Ingeniería del Agua* 18 (1): 99–110. https://doi.org/10.4995/ia.2014.3160.

Rosado, S. 2010. “La gestión territorial y los recursos hídricos en España.” In *Políticas territoriales y tendencias en la administración pública del agua*, edited by J. Mora, F. Dos Reis, and B. Cavaco, 47–88. Lisboa: Instituto Superior de Ciências Sociais e Politicas da Universidade Técnica de Lisboa.

Ruiz, A. 2013. “Reflexiones sobre la gestión de los servicios urbanos del agua: un recorrido histórico del caso español.” *Agua y Territorio / Water and Landscape* 1: 31–40. https://doi.org/10.17561/at.v1i1.1031.

Ruiz, A., and M. A. García. 2017. “Public Participation in European Water Management: From Theory to Practice.” *Water Resources Management* 31: 2479–95. https://doi.org/10.1007/s11269-016-1355-1.

Ruiz, A., M. A. García, and F. González. 2010. “Analysis of Urban Water Management in Historical Perspective: Evidence for the Spanish Case.” *International Journal of Water Resources Development* 26 (4): 653–74. https://doi.org/10.1080/07900627.2010.519497.

Ruiz, A., F. González, and A. J. Picazo. 2015. “The privatisation of urban water services: theory and empirical evidence in the case of Spain.” *Investigaciones Regionales – Journal of Regional Research* 31: 157–74.

Suárez, M. 2020. “Las políticas de tarifas urbanas del agua en España.” *Presupuesto y Gasto Público* 101: 205–26.

Swyngedouw, E. 2007. “Dispossessing H2O: The Contested Terrain of Water Privatisation.” In *Neoliberal Environments: False Promises and Unnatural Consequences*, edited by N. Heynen et al., 51–62. New York and London: Routledge.

Villarroya, F. et al. 2014. “Datos básicos sobre el uso del agua en España: recursos didácticos.” *Enseñanza de las Ciencias de la Tierra* 22 (3): 217–29.

CHAPTER 3

Between Private and Public: Water Supply Infrastructures in Portugal in the First Half of the Twentieth Century

An Attempted Comparison with Spain

Maria Ana Bernardo and Ana Cardoso de Matos

Abstract

The new water supply and sanitation services, considered key elements of city modernisation, were part of the sanitary revolution and a response to criticisms and complaints from urban populations confronted with unhealthy city life. In the framework of the regulatory model, public authorities created the appropriate legal and institutional conditions for the construction and management of these new utilities. In many cases it was they who managed the new infrastructures, either directly or through municipalisation. The purpose of this chapter is to analyse the history of the establishment of modern water supply infrastructures in Portugal in the first half of the twentieth century. We focus on the solutions adopted for the management of water supply networks and the legal and institutional framework that led to them. Finally, based on the same chronology and perspective of analysis, we attempt to compare the Portuguese and Spanish cases.

Keywords

infrastructures – municipalisation – private management – urban modernisation – water supply

1 Introduction

The study of technical networks has received increasing attention from researchers, especially since the 1970s. The most frequent lines of enquiry include the history of technology, business practices, public policy, relations between public and private actors in the construction and management of

 | DOI:10.1163/9789004679634_004

networks, social practices associated with and arising from their installation, and environmental issues associated with their installation and operation (Fettah and Bocquet 2007, 1–13).

In the field of urban studies, these topics have been analysed with the modernisation of cities from the mid-nineteenth century onwards as a focal point. The rapid growth of the urban population and the resulting sanitary, environmental, public health and safety problems put pressure on public authorities, the central state, and local administrations, for greater intervention in urban space. In this context, initiatives emerged for the construction and installation of utility networks, including those related to water and sewerage (Tarr and Dupuy 1988; Lees and Lees 2013, 464–82).

Considered as an axial element of the modernisation of cities, the new utilities related to water supply and sanitation were part of the so-called sanitary revolution and were a field of application of new processes, techniques and materials made available by industry and by the advancement of scientific and engineering knowledge (Konvitz, Rose, and Tarr 1990, 32–37; Vleuten, Oldenziel, and Davids 2017, 29–34). In addition to being a response to the criticism and complaints of urban populations about the difficulties and insecurity of life in cities and materialising the hygienist and social ideals of some sectors of the educated middle classes, modern infrastructures also functioned as devices for inculcating a moral of self-regulation for the inhabitants (Munck 2017).

Many of the instruments of urban planning, control of economic activities and public intervention in the life of cities that emerged since the second half of the nineteenth century resulted from the above-mentioned constraints (Silva and Sousa 2009). Urban development became marked by the so-called regulatory model, characterised by the growing importance of public authorities, from the state to municipalities, in the processes of city building (Newbery 2002, 23–51; Cliftona, Lanthier, and Schroter 2011, 659–72).

The intervention of public authorities was also associated with the technological complexity present in the construction of water and sewage networks. The fact that such networks were installed at the level of the urban subsoil required not only mastery of technical skills in terms of construction techniques and materials used, but also knowledge of the topography and geology of the places where they were to be installed and the planning of their layout. The knowledge of engineers became indispensable both for the design of projects and for monitoring the work of installing them (Fernández-Paradas and Matés-Barco 2022, 75–111; Cardoso de Matos 2019, 13–22). But the intervention of the public authorities in the construction of these infrastructures was not only due to their technological complexity. It was also determined by political, economic, and legal factors, and by the relationships established between the different actors in the process. As Fougères (1998, 202–3) states, the home water supply service is, like other urban infrastructures, a *lieu d'arbitrage* between what is sociopolitical and what is geotechnical.

In Portugal, the issue of public or private management of water supply infrastructures emerged during the second half of the nineteenth century, concomitant with the start-up and installation of the same infrastructures in some cities (Silva 2020, 95–9; Silva and Cardoso de Matos 2004, 1–30). It was then considered, as in other countries, that the best solution was private management, either through ownership or concession extended to companies, usually in a natural monopoly regime (Matés-Barco 2004, 165–77; Milward 2000, 315–50).

In this chapter we intend to analyse the dynamics of the construction and management of modern water supply infrastructures in Portugal in the first half of the twentieth century. Based on the Portuguese case, an attempt is made to compare it with the historical experience of Spain, with special focus on Andalusia. The chronology chosen starts in the mid-nineteenth century, when, as stated above, the construction of modern urban infrastructures began. At the end of the analysis, we considered the period that has been pointed out as that in which, from the management point of view, municipal solutions gained relevance in relation to entrepreneurial ones.

Regarding Spain, the 1940s corresponds to the period from which this turnaround occurred (Matés-Barco 2021b, 135–59). In relation to Portugal, the historiography on the subject does not allow for the same inferences, although it is noted that only in the case of Lisbon was management always carried out by a private company, within a framework of marked regulation, natural monopoly and by means of legal conditions that allowed the viability of exploitation by the company (Silva 2020; Saraiva, Schmidt, and Pato 2014). The information available for the rest of the country, in the cases where a private company was involved in the construction and/or management of water supply, refers to the 1930s as the turning point for the public management of this urban infrastructure, through the option of municipalisation (Silva and Cardoso de Matos 2004).

Despite these differences, we believe that the hypothesis of their resulting from institutional practices stemming from different political-ideological opinions on public or private management of water supply infrastructures is not very operative. Studies on the history of the construction and management of water supply infrastructures in Portugal are still scarce and are mainly of a monographic and descriptive nature, as the cases of Porto (Amorim and Pinto 2001), Braga (Cordeiro 2018), Figueira da Foz (Brandão and Callapez 2017), Coimbra (Mendes 2013), Lisbon (Pinto 1989; Silva 1997b) and Evora (Bernardo 2004).

Thus, subject to the need for greater empirical evidence, our line of enquiry is based on the hypothesis that the construction and management of modern water supply infrastructures acquire greater intelligibility if understood in the framework of the "urban question", and its specificity in the Portuguese case. In the period under analysis, the Portuguese urban hierarchy and network was characterised by the macrocephaly of Lisbon, by the marked population distance of Porto, the country's second city, in relation to Lisbon, by the scarcity of medium-sized cities, whose population growth was not very expressive, and by a larger number of small cities, which remained at that level (Silva 1997a, 779–814; Nunes 1996, 7–47; Veiga 2003, 91–109). One of the dimensions of the so-called "urban question", as a response to the population and health pressure on cities, concerns the growing involvement of public authorities, from the second half of the nineteenth century, both in the production of regulatory laws and in directing the construction and management of network utilities, among them those related to water supply (Millward 2005, 33–58).

In Portugal, urban population pressure and the resulting problems in terms of hygiene and public health, except in the cases of Lisbon and Porto, did not have the same magnitude as in other European cities. However, urban leaders, especially those in charge of medium-sized cities on the scale of the Portuguese urban network, appropriated the arguments of technological modernisation and used them in discourses of competition between cities and to legitimise their actions as local governors. When private capital and companies did not correspond to the appeals of the public authorities for the installation of utilities infrastructures, the decisions of direct management or the option of municipalisation did not necessarily derive from convictions in matters of economic policy (Bernardo 2013, 281–313). However the political and the symbolic were not alien to decision-making and could function autonomously in the face of economic rationality. Additionally, in Portugal, with the establishment of the dictatorial Estado Novo (New State) regime in the early 1930s, those arguments were combined with public policies of an interventionist nature, which favoured the predominance of municipal water supply management in urban areas.

The first section of the chapter follows the initial stage of the construction of modern infrastructures in Portugal. The analysis is primarily based on the institutional arguments and the legal framework that was designed and highlights the period and the arguments from which the municipal option became predominant regarding the management of modern water supply infrastructure in urban areas. In the second section of the chapter, an attempt is made to compare Portugal and Spain, based on data from specialised literature and taking into account, above all, the political and institutional aspects of the process.

2 The Institutional and Legal Framework for Constructing Modern Infrastructures

2.1 *From the End of the Nineteenth Century to the End of the 1920s*

Matters regarding lighting, water supply and sanitation in urban centres figure among the main responsibilities assigned to local authorities and were systematised in the administrative codes introduced in Portugal following the establishment of the monarchical Liberalist government.

The last Code of the monarchical period (1896), which retained the discursive formulas of the previous administrative legislation, regarded the administration of the property and affairs of *concelhos* (municipalities) as fundamental competencies and responsibilities of local councils, as well as the planning and implementation of measures for material improvement with a view to improving the quality of life of local people (Código Administrativo de 1896, 23–29). Also in Spain, at the beginning of the twentieth century, the Municipal Law of 2 October 1877 was still in force. This law, which remained in force until the publication of the Municipal Statute of 1924, established that municipalities were responsible for various local services, including lighting, water supply and sanitation. Both the Portuguese and the Spanish municipalities directly exploited the services that required fewer financial resources, such as public cleaning, cemeteries, abattoirs, and markets, and awarded private concessions to exploit the services that required greater financial and technological resources, such as water supply, lighting and sanitation. In the case of Portugal, the Administrative Code considered that municipalities were responsible for the maintenance work and improvements to facilities and systems which traditionally provided local residents in urban centres with access to water and lighting, and it also covered studies and plans for infrastructures of this nature, drawn up in accordance with modern engineering concepts.

For the period considered in this analysis, in the case of water supply it meant the carrying out of periodic scientific analyses for quality control for domestic use, new sources of energy and machinery for the purposes of harnessing water at source, distribution and storage, and the construction of domestic supply networks for public and private consumption. Thus, the new standards of comfort and hygiene provided by the technological and scientific progress characteristic of industrial societies became generalised and were much sought after by the urban middle classes.

Whether to pursue a traditional policy of material improvement or the option for innovatory technical solutions was up to municipal councils. Such decisions depended, in the final analysis, on the knowledge and competencies of the elites who led council administrations and how receptive they were to innovation. Case studies carried out for Portugal seem to point to a positive correlation between the pursuance of modernising policies and the presence of engineers, who either worked for local authorities or frequented the same circles of sociability as the individuals who made up council teams.

Financing was necessary for construction projects and very often placed an insurmountable obstacle in the way of implementing modernising projects by local authorities in Portugal. Funding such investment projects derived from municipal income from council assets, taxes levied, and bank loans contracted. As most Portuguese municipalities councils suffered from a chronic lack of funds, borrowing often assumed worrying proportions and led to burdens of debt payments that could threaten councils' financial viability. The Administrative Code of 1896 sought to minimise this problem by placing tighter controls on borrowing by local councils and increasing central government control over municipal authorities (Serra 1988, 1037–66). The new Code was criticised as aiming at excessive centralisation and representing an attack on the traditional autonomy of Portuguese municipalities. The Republicans used this argument in order to attack monarchical government departments and promised to introduce a more decentralised model of administration under a future Republican regime. But the circumstances resulting from the increase in the number of people living in urban centres in Portugal led the state, as early as the period of the Liberal Monarchy, to introduce legislation designed to provide a legal framework for intervention aimed at improving conditions of sanitation and hygiene in the main cities and towns in the country.

The outbreak of bubonic plague which hit the city of Porto in 1898 was instrumental in the setting up of the *Conselho de Melhoramentos Sanitários* (Sanitary Improvement Council) in 1901, under the *Direcção Geral das Obras Públicas e Minas* (Department of Public Works and Mines) (Decree 14/10/1901). The *Regulamento Geral de Saúde e Beneficência Pública* (General Regulations for Health and Public Welfare), dated 24 December 1901, aimed at providing a framework for local authority intervention in the field of public health, were also introduced in 1901 (Chart of Law 24/6/1901). This outbreak of bubonic plague in the city of Porto was directly related to the decision of the Spanish government, dated 1899, to create the *Dirección General de Sanidad* (Health Directorate General) (Rodríguez 1994, 20–1). In 1903 this directorate was replaced by the *Instrucción General de Salud* (General Health Instruction) which created two General Inspectorates, one for the inspection of internal health and the other for the inspection of external health.

In Portugal, a move was made in 1903 to improve the habitability of urban buildings with the introduction of the *Regulamento de Salubridade das Edificações Urbanas* (Urban Buildings Health Regulations) (Decree of 14/2/1903), which laid down guidelines to be followed in the construction of urban buildings. These stipulated not only that council licences for construction or rebuilding works should be subject to prior approval by regional departments of the Sanitary Improvement Council but also made it compulsory for builders to comply with the provisions of the *Regulamento Geral de Saúde e Beneficência Pública* (General Regulations for Health and Public Welfare).

In an attempt to gauge the situation on the ground in the main cities and towns of the country, in 1903 the government ordered a public health survey to be carried out in urban centres (Montenegro 1903). Also in Spain, in 1908 the *Sociedad Central de Arquitectos* (Central Society of Architects) published a report on the construction of cheap and hygienic houses. However, the measures that made it compulsory to introduce hygienic conditions in dwellings were not immediately implemented. An important moment for the introduction of these measures in the various municipalities was the approval, in 1913, of the *Reglamento para la Higiene y Salubridad de la Ciudad Alicante* (Regulation of Hygiene and Salubrity of the City of Alicante), and the following years were marked by the emergence of various studies and proposals on this subject (Pérez-del-Hoyo and Gutiérrez-Mozo 2013, 223 and 234).

The revolution of 5 October 1910 ushered in the Liberal Republic. Although Republican leaders intended to quickly draw up a new Code, which faithfully reflected their ideas as regards administrative decentralisation, they ended up by maintaining in force part of the Code of 1896, while introducing certain new guidelines of a decentralising character in the administrative Code of 1878. As regards the question of material improvement, the most relevant aspects of these changes were the return to a greater degree of flexibility towards borrowing in order to finance public works and to less stringent control by central government. The execution of studies and the drawing up of plans for the construction of modern urban water-supply and lighting infrastructures continued, therefore, to be one of the main responsibilities assigned to local authorities and was still carried out on a casual basis.

Nevertheless, Republican decentralisation motives are present in the administrative legislation of 1913 (Law no. 88 of 7 August 1913), especially the provision for municipalities to form groups in order to carry out certain improvement works where the volume of investment was beyond individual councils' capabilities. There was also provision in the bill for endowing local authorities with the power to discuss the "*municipalisation*, or local authority provision, of services to local people", but in the event this did not become law. Only in 1927, after the Municipal Congress of 1922 and the Third Electricity Congress of 1926, at both of which the need for councils to take charge of provision of services was reaffirmed, was this provision granted legal authority

(Decree no. 13350 of 25 March 1927).[1] The regulations contained in the law stipulated that municipal services should be provided by departments retaining autonomy within the framework of local authority administration and that they should be managed by an administrative committee elected by the council. Each department responsible for providing municipal services should have a team of technical, administrative, and manual staff to ensure the smooth administration of the system and planning and execution of works. Councils were also given specific powers to manage services and assets in the following areas: the construction and operation of water-supply systems for domestic consumption; cleaning, sanitation and public sewers, including the removal or use of urban waste; the production and distribution of gas and electricity; the construction and operation of markets, abattoirs and refrigerators; the construction and operation of public baths and washing sites; the construction and operation of means of transport; and any other services involving the production or distribution of goods or the provision of services aimed at satisfying the needs and interests of local people. At the same time, there was a provision to the effect that when a need for the operation or development of certain services required loans to be contracted, the municipal services administrative committee should not only justify the need for borrowing but also guarantee to fulfil the terms of repayment.

The law of 7 August 1913 also enabled local councils to form federations in order to jointly administrate one or more types of public utility. Given that setting up departments to provide urban infrastructures required a huge financial investment, the purpose of this measure was to ensure the best use of such investment by councils. In fact, the concern of the government to encourage the provision of municipal services led to the gradual introduction of a series of laws during the subsequent period. Councils, which operated directly and were fully financially responsible for providing supplying water, public lighting and traction services and gas and electric lighting and energy to private consumers, were also granted exemption from *contribuição industrial* (industrial contribution) tax. In 1920, in order to encourage municipalities to start providing water and lighting to households, the government exempted them from import taxes on all imported materials needed for the construction of water supply and lighting networks and simplified the legal procedures for the expropriations needed to establish these urban networks (Decree-Law no. 1789 of 29 June 1925).

1 The regulation of the provisions of this decree took place on 27 July of the same year.

By 1927 it was recognised that there was a significant degree of municipal involvement in the provision of certain public utilities in the country as a whole (Decree no. 13350 of 25 March 1927), although only around 15% of councils had their public utilities municipalised. A clear example of the influence of the administrative decentralisation ideas dear to the Republicans is provided by the "municipalist" movement of the early twentieth century, which held congresses across the country. Seen as championing the cause of the liberty enjoyed by the districts in the past, the movement advocated the free association of municipalities and the formation of a federation with the aim of strengthening local power.

For example, at the *Congresso Municipalista Alentejano* (Alentejo Municipalities Congress) held in 1915, those attending backed the modernisation of Alentejo municipalities for the "development of the Alentejo" by pushing for increased autonomy with regard to taxes, infrastructures, economic activity, and the spread of farming techniques and assistance, culminating in the setting up of a *Federação de Municípios Alentejanos* (Federation of Alentejo Municipalities) which would lead to the establishment of a provincial parliament for the discussion and deliberation of matters which were of interest to the Alentejo (Bernardo 2004). However, such proposals were never put into practice either in the Alentejo or in any other region, although they did faithfully reflect the plan for administrative decentralisation put forward by the Republicans.

In order that plans for federations of municipalities and provincial parliaments could be carried out, a major change was needed in the model of local authority administration which had existed in Portugal for around a century. But the Great War forced the Portuguese Republican regime to concentrate on the management of the political and military stance of the country regarding the conflict and the resolution of economic and social problems brought by the war. After 1918, there was increasing political instability and critics of the regime gained ground.

In Spain, the publication of the Municipal Statute in 1924, the drafting of which was directly linked to the jurist Calvo Sotelo, who in December 1923 had been appointed general director of administration by the Primo Rivera dictatorship, was an essential measure for the legal, administrative, political, and urban organisation of Spanish municipalities (Faria 2015, 107). On 12 October 1925, the idea arose of organising a municipal association movement in Spain. Municipal associations were presented at the First National Municipalist Congress held in Madrid, which was attended by 429 delegates representing 85 municipalities. Once this idea was well accepted by the delegates present, it was agreed that it would be presented to the representatives of the other

Spanish municipalities in order to be formalised during the Second National Municipalist Congress which took place in 1926. Thus, in this congress the *Unión de Municipios Españoles* – UME (Union of Spanish Municipalities) was officially created, and its statutes were approved (Orduña 2006, 335). The main objective of the UME was to study the "great problems posed daily by the administration of the city and the organisation and operation of its services", concepts that were undoubtedly modern (Orduña 2006, 337). In 1931, at a time when a republican regime was in force and an environment of public freedoms was being experienced, a new internal regulation of UME was approved, replacing the regulation of 1926, and introducing various democratising modifications to its structure. This new regulation provided for the possibility of local authorities joining UME individually or by forming regional, provincial or district federations (Orduña 2006, 355).

2.2 *Municipal Management from the 1930s Onwards*

In 1926, a civilian-backed military coup installed the Military Dictatorship, thereby putting an end to the Parliamentary Republic in Portugal. At the administrative level, this situation removed any possibility of achieving the decentralising aims of Liberal Republicanism, even more so as the years passed. António de Oliveira Salazar emerged on the Portuguese political scene in 1928 and was responsible for shaping the development of the country over a period of several decades; in 1933, under his direction, a repressive dictatorial political regime was established which was known as the *Estado Novo*.

District councils became vehicles for the implementation at the local level of government policy, which prejudiced their status as municipal administrative bodies representing the interests of local people, the purpose for which they had been elected. The Administrative Code of the New State, which became law in 1936, served to confirm the subsidiary role of local authorities.

From the beginning of the 1930s, the state was concerned not only with bringing economic activity as a whole within the legal framework but also, and increasingly evidently, with following a policy of state direction and control (Rosas 1986, 254–83). The economic and social effects of the crisis of 1929 facilitated this stance. From at least 1931 the government sought to officially gauge unemployment in Portugal, and in 1932 legislation was introduced which made a concerted effort towards resolving the problem (Decrees no. 20979 and 20980 of 7 March 1932). In the first quarter of that year the *Alta Comissão para Melhoramentos Públicos* (High Commission for Public Improvements) was set up under the Ministries of Finance, Commerce and Communications, and Agriculture and the government took measures for the exceptional provision of funds for public works. Some were targeted at the so-called rural improvements and were to be distributed to almost all parishes and districts in the country; while funds were also channelled into the "great cities" with a view to launching and implementing public works considered to be "of the greatest social utility", such as council housing estates in Lisbon, further education colleges, a hospital in Porto and a lunatic asylum in Coimbra.

In July of the same year, Salazar took over as Chief of Government, the first civilian to occupy the post since the military coup of 1926. Two days later the Ministry of Public Works and Communications was set up, replacing the Ministry of Commerce and Communications. The symbolism of this change was evident: the legacy of the Liberal Republic was banished and at the same time the designation of the old Ministry of Public Works, Commerce and Industry, which the national collective memory associated with the intentional policy of material improvement carried out by the monarchical regime in the second half of the nineteenth century, was revived. Funds earmarked for unemployment benefit were channelled into the new ministry and a stage of the New State was initiated in which public works programmes gained in importance and visibility (Tostões 1996, 585–98).

Also on 19 September 1932, several laws were passed which together established the legal framework for the policy of material improvement and created the mechanism for state direction of the process as a whole. Each law dealt with a different area of actuation and defined the role to be played by central government and determined what kind of control it would exercise in each area (Decrees no. 21696, 21697, 21698 and 21699, all approved on 30 September 1932).

Rural improvements were defined as works to be carried out at the local level outside urban centres and district capitals, such as: highways, local tracks, road surfaces, fountains, water tanks and washing sites. Although the government considered that such works "could involve a degree of decentralisation by means of cooperation with local authorities", it added that local councils "could only achieve greater efficiency through technical centralisation". This role was entrusted to the *Junta Autónoma de Estradas* (Autonomous Highways Board), made responsible for assessing and approving plans presented annually by local authorities. On the basis of these plans, the Ministry of Public Works would provide funding to a maximum of fifty per cent of the total cost of works, while councils would have to find the remainder.

Urban improvements were defined as those "which brought the greatest benefit in terms of the welfare of people living", which the government, "as coordinator and director of all the activities of the nation sought to carry out on a general scale in order to foster the development of villages, towns and cities in the country" (Decree no. 21698 of 30 September 1932). Works regarded as urban improvements were "those of interest at the local level which would benefit the community, carried out outside large population centres, including the drawing up of urban development plans, the construction, repair and remodelling of primary schools, elementary professional schools, municipal secondary schools, hospitals and other buildings with a social role, museums and national monuments". Carrying out such improvements required the drawing up of "a general plan whereby state aid to local people, either by direct means or through intermediaries such as state departments and local authorities, might ensure that such improvements are executed in accordance with effective guidelines". The *Direcção Geral dos Edifícios e Monumentos Nacionais* (Department of Buildings and National Monuments) was charged with providing technical guidance and inspection services and conducting surveys in order to assess local needs, while the Ministry of Public Works was assigned responsibility for the distribution of funds and overall control of procedures, and there was provision for the sharing of the burden of investment between the state and local authorities.

Improvements to waterworks and sewers were defined as "works to harness at source and distribute water and the construction, improvement and extension of sewer networks in towns and important settlements and in cities, except in large urban centres". The "very valuable contribution" made by local authorities in this field was acknowledged and "the huge dimension of the problem" was recognised, while the bill made it clear that "the state had a duty to facilitate the mission of these bodies and ensure that it was carried out" (Decree no. 21698 of 30 September 1932).

The Department of Health was given five months in which to carry out a survey of

> district capitals, main towns in districts and other important towns and villages in each district, providing details of the potential of each type of public service and the weaknesses identified in services in operation that should be remedied, highlighting the most urgent cases in each *concelho* (municipality) and *distrito* (district). (Decree no. 21698 of 30 September 1932)

Once the results had been obtained, the government would be responsible for drawing up a programme of improvements to be carried out by councils and ordering the execution of studies and the drawing up of plans through the competent departments. The *Administração Geral dos Serviços Hidráulicos e Eléctricos* (Department of Hydraulic and Electric Works), under the Ministry of the Public Works, could authorise local authorities whenever it deemed necessary to contract out the execution of studies and plans to third parties with technical expertise. The law also provided for the programme of improvements to be reviewed every three years to take account of the results of new surveys; the *Administração Geral dos Serviços Hidráulicos e Eléctricos* would submit a report on the plan of works to be carried out by district councils by 31 March of each year and, having been approved by the *Conselho Superior de Higiene e Obras Públicas* (Council for Hygiene and Public Works), plans would be submitted for approval by the Ministry of Public works. The cost of these improvements was to be borne by municipal councils, except where this was recognised as being impracticable, in which case the state would share the cost of investment up to a maximum of fifty per cent of the cost of works. Specific provision was also made for county councils, committees responsible for proposing plans and private individuals to be able to carry out improvements on their own account, with the technical assistance of the state, provided that such works had the prior approval of the *Administração Geral dos Serviços Hidráulicos e Eléctricos* and were included in the plan of works approved by the government.

Also on 19 September 1932, in order to combat the employment crisis affecting the country, the government set up the *Comissariado de Desemprego* (Unemployment Commissariat), with departments in each county. This was a department of the Ministry of Public Works, although its operations and administration were autonomous. Its main aims were to organise the registering of the unemployed, help unemployed people to find work and administrate the system of unemployment benefit. Among works which would most profitably benefit from the harnessing of this source of labour, the government regarded "rural improvement works, urban improvement works, water and sanitation works, the cleaning of ditches and verges, tree-planting, and so on". Among works to be carried out in order to relieve unemployment, the law classified water-supply and sanitation infrastructures as complex works requiring a great degree of technical skill and therefore those which "will take a considerable time to become fully operational over extensive areas, although due to their great value in terms of bringing improvements in urban hygiene, they are works whose execution is most necessary"; the close relationship between the rate of execution of sanitation works and the decrease in death rates is evident here. Taken together, these laws represented a blueprint for concerted intervention by the government in the field of public works and identified the New State with the values of modernisation and material progress. The improvement programme thus involved a discourse that, by valuing "work accomplished", aimed at legitimising the authoritarian regime and portraying the Liberal Republic as a period of anarchy and political uncertainty.

In addition, in late 1934, a law was passed requiring all towns and cities which were district capitals to draw up topographical plans and urban development plans for settlements with a population of over 2,500 with a high rate of demographic growth and for centres or zones with interest of a tourist, recreational, climatic, therapeutic, spiritual, historical or artistic nature, as designated by the government (Decree-Law no. 24802 of 21 December 1934). As justification for this compulsory requirement, the government cited a combination of "the benefit of social assistance through providing work" to local people who needed it with the fact that district councils requested "the intervention of the government" for plans which were not always "drawn up according to the best criteria and the fairest consideration of local conditions and future needs, or in compliance with the best standards of hygiene and convenience for urban settlements". It also referred to the frequent lack of "technical staff with specialised knowledge of the modern art of urban planning, which is also a science".

The drawing up of topographical plans, with the exception of those of Lisbon and Porto, was offered for public tender, based on criteria established by the government, which was also responsible for the overall direction and inspection of works. However, councils were expected to reimburse the state in instalments for expenses incurred. The law also provided councils with a three-year deadline following the drawing up of topographical plans to present their individual urban development plans, which for government approval had to be granted following the issue of a certificate by one of the *Conselhos Superiores de Obras Públicas, de Higiene e Turismo* (Higher Councils of Public Works, Hygiene and Tourism). Among the compulsory requirements for urban development plans were schematic plans of existing water-supply and public lighting networks, water-drainage systems, and sewer systems, along with a statement of what was required to be done in this field. In theory, the new legislation provided for new zones of growth and expansion of urban centres to be endowed with modern lighting and water-supply and sanitation infrastructures.

All the legislation examined above shows an aim on the part of the political leaders of the New State to restrict the scope of local authority initiative in the field of infrastructures and urban facilities by means of instruments of control such as guidelines, technical inspection and the financial control of plans. In certain cases, the Ministry of the Public Works directed works itself, through central departments and/or regional and other departments. In fact, the intervention of the government in these areas was based on a broader aim than that which was contained in the provisions of the law of 1933, which was approved by the *Estatuto do Trabalho Nacional* (National Work Statute). The law stated that "the state has the right and the obligation to coordinate and regulate from above the economic and social sphere and set objectives for it" (Decree-Law no. 23048 of 23 September 1933).

In accordance with this guideline, the *Lei de Reconstituição Económica* (Law of Economic Reconstitution) was introduced in 1935, providing for "plans and works for whose execution there is a pressing need over the next fifteen years", and earmarking substantial funds, considered as extraordinary expenditure (Law no. 1914 of 24 May 1935). Resources deriving from the success of financial and monetary stabilisation policy carried out in previous years would be channelled into the modernisation of the armed forces and the improvement and/or construction of basic infrastructures for the economic development of the country. Areas identified for investment were: roads, railways, ports, airports, telegraph and telephone systems, electrification, agricultural hydraulics, public buildings and monuments and a range of other urban facilities in Lisbon and Porto, as well as colonial credit.

In view of the ambitious nature of such proposals, when the bill was debated in parliament, the government made it clear that it did not seek to follow the development policy pursued in Portugal between the 1850s and 1880s, nor did it have any affinity with the socialist planning model (Nunes and Valério 1996, 512–3). However, when extensive legislation was introduced in 1944 dealing with the question of water supply and sanitation, the state decreed that the construction of drinking water and sewer networks in district capitals in continental Portugal was a *mandatory issue and a question of priority*, overriding all others (Decree-Law no. 33863 of 15 August 1944). In the preamble to the law, a brief outline was given of the development of the country as regards improvement works, highlighting the "extremely grave situation" in which the nation found itself. Although it was pointed out that from 1932 the state had taken measures "of the utmost importance" for the sanitary state of the country, it was evident that there was still much to be done. As such works were of great complexity, and carrying them out was not mandatory by law, many local authorities had decided to build monuments or carry out urban development works instead of executing works to construct water-supply and sanitation systems.

For this reason, while declaring that it could not resolve the problem with a single programme, the government proposed to provide drinking water to all towns and cities which were district capitals by carrying out a plan over a ten-year period: a joint action involving the state and municipal councils, although with a "completely new feature: the mandatory character of the execution of such improvements". From the technical point of view, it was decided that local councils would continue to be responsible for carrying out studies and executing works, although there was provision for the state to take the initiative for construction projects and/or intervene in operations to carry out works in certain cases. Works would be executed in accordance with an annual execution plan drawn up by the Ministry of Public Works and Communications and approved by the Ministers of Public Works and the Interior. The Department of Hydraulic Works would be responsible for organising the tendering of contracts for studies and plans to be carried out annually.

The financial feasibility of the undertaking was assured in the following way: councils were given the power to contract loans up to the value of fifty percent of the total cost of works, and offered favourable conditions of repayment, while in order to safeguard the councils' regular income, loans were to be guaranteed by income from the sale of water; to cover the remaining investment cost, a subsidy of a minimum of twenty-five per cent of the total debt burden was to be provided by the Unemployment Benefit Fund, and a central government subsidy would also be provided, if required, "out of its general income".

As regards the management of these urban utilities, councils were given the option of awarding contracts or operating the services themselves, but with separate accounting from the rest of the municipal finances. Only in exceptional cases, where council income was too low to allow for either of these options, did the state authorise the direct management of services by councils themselves. Thus, an attempt was made to remedy situations in which the lack of sufficient income for the use of the new infrastructures might jeopardise municipal spending.

The law of 1944 shows clearly that the intention of the state was to control the process of the building of modern water-supply and sanitation infrastructures while recognising that in the mid-twentieth century, the situation of Portugal in this field still left a lot to be desired. While based on ideological grounds, this obligatory approach was also the result of the recognised incapacity of municipalities to satisfactorily meet needs for the improvement of standards of hygiene, health and comfort for people living in towns and cities around the country.

In 1927, the year after the military coup that established the Military Dictatorship in Portugal, jurist Luiz Jordana de Pozas, speaking at the Spanish conference *Intermunicipalismo, Mancomunidade y Asociaciones de Municipios* (Intermunicipalism, Mancomunidade and Associations of Municipalities), defended *intermunicipal* (intercity) cooperation, stating that "I understand intermunicipalism as the joint and spontaneous action of several municipalities for strictly municipal purposes [...] which is built on the respected basis of self-government" (Pozas 1961, 686, quoted by Faria 2015, 117). Also in 1927, the Third Municipalist Congress was held in Barcelona. At that time in Spain, people still believed in the autonomy of the municipalities, but they were aware that without cooperation between them it could not be achieved. The year 1929 was marked by the UME's approach to international municipalism. In that year the IV Fourth International Congress of Cities and Local Bodies of the International Union of Local Authorities – IULA, was organised in Seville. This congress marked the "institutional and professional consolidation of Spanish municipalist thought on the international scene" (Faria 2015, 118).

During the years of the Spanish Civil War, the UME was to have much less activity. After the end of the war, the directive council of this organisation still met in Madrid, but in September 1940 a law was published extinguishing the UME and creating the *Instituto Estudios de Administración Locales* (Institute for Local Administration Studies), whose main functions were research, study, teaching and propaganda in matters related to local administration and which functioned as an advisor to the government on these matters, which shows a clear intention to centralise the affairs of municipal administration (Orduña 2006, 359).

3 Infrastructures in the First Half of the Twentieth Century: A Comparative Perspective from Portugal and Spain in Water Supply

An assessment of the achievements regarding the construction of water-supply networks in Portuguese towns and cities is complicated by the absence of statistics dealing with these matters before the 1930s. As far as water supply is concerned, the first systematic survey of the nationwide situation was conducted in 1934. This initiative was prompted by a desire on the part of the government to find out the current state of the water-supply and sanitation system in the country following the legislation of September 1932.

It could be speculated that conducting such surveys was not regarded as a matter of priority, as for many years central government regarded the creation of these urban infrastructures as the responsibility of local authorities. Even if there is some truth in this, this scenario resulted above all from the difficulties central government encountered in endowing local authorities with efficacious means for obtaining the necessary information (Sousa 1995, 250–3). The putting in place of modern infrastructures for the harnessing of water at source and distribution advanced at a snail's pace. The information available, based on the above-mentioned survey of 1934, covers all towns and cities in continental Portugal which were district capitals, with the exception of the cities of Lisbon and Porto; the authorities decided that separate surveys should be carried out for these distinct cases. The results of the survey enable a distinction to be made between the harnessing at source and supply of drinking water to public access points in urban centres, and the construction of water distribution networks to public supply points and to homes. The former involved quite extensive improvement works, both to collection points and pipes for carrying water to consumer areas, where public springs and fountains were built. In many cases, existing facilities already cortar meet the needs of the growing population of the urban centres they served. In other cases, works were required to allow water to be carried from places some distance away from towns and cities, and recourse to the latest techniques and technology was required in order to negotiate natural obstacles. The survey covered two hundred and seventy district capitals, with the following distribution in terms of the resident population: 75% of these towns and cities had fewer than 5,000 inhabitants; 16% had more than 5,000; 7% had more than 10,000; and only 2% had a population of over 20,000. That is, around three quarters of settlements surveyed had populations that were below the quantitative threshold for designation as urban centres. However, from the administrative point of view, they were all district capitals, and as such they were the district administrative and political power base.

The characteristics of the Portuguese urban network placed limits on the construction of modern networks for harnessing water at source and supplying it. Firstly, the small size of the majority of population centres restricted municipalities' capacity for generating income and forced them to borrow money to finance expensive investment projects. Moreover, private investors and companies looking to construct and operate such infrastructures were less likely to be interested in small towns and cities, given the importance of the economies of scale required to make such enterprises profitable.

In settlements with up to 10,000 inhabitants, more than 50% of water sources were located less than a kilometre from the urban centre in question. The survey results show that in many cases water was sourced within or just outside the boundaries of the settlements. Such sources included springs, wells, artesian wells and rivers, in which case relatively simple works and plumbing arrangements were sufficient to provide local people with access to water. In urban centres with a large population, water sources tended to be located further away. In towns and cities with a population of over 10,000 inhabitants, approximately 17% of sources were located more than 5 km away from settlements, and in centres with over 20,000 inhabitants the percentage was around 33%. In such cases, works to harness water at source were complex and involved a considerable investment. It should also be noted that in 45% of the total number of cases surveyed, the rate of flow of water at source was unknown, and for others it was considered insufficient for consumption needs. Thus, the conclusion can be drawn that the situation in which the country found itself with regard to infrastructures for harnessing water at source was still very backward in the mid-1930s and this certainly restricted the opportunities for constructing or extending water distribution networks.

Of the two hundred and seventy district capitals surveyed, only 36% had domestic water supply distribution networks in place. And in these cases, it should be noted that only a very small proportion of local people enjoyed the benefits of the system at home: in around 15% of settlements, water was only piped to public springs and fountains, and for 48% of settlements there is no mention of the existence of a distribution network. Access to water for the overwhelming majority of people living in the urban centres surveyed meant getting supplies from public or private wells, springs and fountains and using the services of water carriers, wherever household incomes allowed.

By examining the number of cases in which water is specifically mentioned as being distributed to public supply points, an assessment can be made of the rate of construction of distribution networks. It is evident that the process accelerated during the 1920s and 1930s: around 16% of networks had been constructed before 1900 or were classified as "old"; in the first decade of the twentieth century about 4% of the total of networks which can be dated were built; in the 1920s, 10% were put in place; and in the following fifteen years leading up to the date of the survey, 53% of networks were built. In fact, 50% of water-supply networks whose construction can be dated were built between 1926 and 1934, a period which coincides with the beginning of the dictatorship and the initial years of the New State. In 1930 about 60% of the 178 Spanish cities with more than 10,000 inhabitants had some kind of water supply company. There was a slight variation in these companies in the following years: 1931 and 1931, 117 companies; 1933, 122 companies; 1934, 120 (Matés-Barco 2021b, 142–3). In Spain in 1933, of the various companies operating water supply networks, 17 were large companies, while the number of medium-sized companies was 44 and the number of small companies 33. As Juan Manuel Matés-Barco states "Catalonia, Andalusia, Valencia, and Murcia were the regions with the highest concentration of companies, both in terms of the number of companies and the amount of capital invested" (2021a, 95).

At that time, in the whole of Andalusia, only in the city of Seville was there one large company, the *Seville Waterworks*, which had been founded in 1881, while the number of medium-sized companies operating water supply in Andalusia was 10, or 25%. The first of these companies, *Aguas de Gévora*, started in Badajoz in 1878, and from then until 1933 the following companies were established in Andalusia: in Cadiz, *Aguas Potables de Cádiz*, in 1885; in Cordoba, *Aguas Potables de Córdoba*, in 1891; in Linares, *Aguas de Linares*, in 1908; in Bujalance, *La Alameda* in 1909; in Martos, *La Fraternidad* in 1900; in

Algeciras, *Andalusia Water C. Ltd.* in 1912; in Villa del Rio, the company *La Estrella* in 1918; in Espejo, the company *La Constancia* in 1923; and in Huelva, the company *Aguas de Huelva* in 1925 (Matés-Barco 2021a, 87–91).

4 Conclusions

From the second half of the nineteenth century onwards, modern urban network utilities were publicly valued for their benefits in solving some of the main urban problems, such as hygiene and public health. In Portugal, regarding water-supply utilities, the state and municipalities also highlighted these advantages, even in urban centres of smaller size and growth, where the population pressure was not so intense. The positive externalities derived from modern water infrastructures (Silva 2020) were recognised by the public authorities and justified the need for their installation, even when from the point of view of raising funds for investment the context was adverse. What is known about the process of installation of the water supply and sewage network in the city of Évora, namely, illustrates precisely this situation (Bernardo 2013 and 2019).

Although, as previously mentioned, the Portuguese case requires further investigation, the comparison between the historical experiences of Portugal and Spain shows some differences in the management of modern water supply networks. In Portugal, the ownership and management, direct or municipal, of those utilities was the most common path for the installation of these infrastructures, through a trend that originated in the second half of the nineteenth century and continued in the first half of the twentieth century. In this country, the water supply companies that were created did not occupy an important place in the Portuguese business sector (Silva 2020; Silva and Cardoso de Matos 2004). In Spain, until the 1940s, water supply companies were ranked fifteenth among private companies (Matés-Barco 2004 and 2021a) and were mainly located in medium and large cities.

The greater urban and industrial dynamism of Spain, when compared to Portugal, is therefore a key element in understanding the differences identified between the two countries regarding the creation of companies linked to water-supply management. As water-supply infrastructures are territorial, and scale and capital intensive, projects associated with smaller urban realities had more difficulty in attracting companies. Once tenders were launched, they were often unsuccessful. This situation slowed down the materialisation of the projects and, concerning Portugal, it was a relevant factor for the preponderance of the municipalities in the management of those projects, with recourse to indebtedness to financial entities. Once the process started, and in more extreme situations, a very relevant part of the municipal resources ended up being channelled towards the payment of interest on the debt (Bernardo 2013). Thus, regarding Portugal, we maintain that the reasons associated with the historical relevance of public authorities in the implementation and management of the water supply infrastructures are the following: the urban network had great specificity, with the macrocephaly of the capital, Lisbon; the scarcity of medium-sized cities and a large number of small urban centres, whose economic dynamism left no room for great expectations regarding the universe of potential consumers, private or public. The very technical complexity of water supply projects, and their close correlation with the characteristics of the territory (water availability, orography, urban morphology), as well as the notion that these projects had an expectation of dilated return over time, are also factors that contribute to explaining the irrelevance of private companies in the construction and management of modern water supply infrastructures in Portugal throughout the period studied.

The growing interest for the option of municipalised management of modern urban infrastructures in Portugal, from the beginning of the second decade of the twentieth century, cannot be dissociated from the republicans coming to power in 1910. In terms of territorial administration, the country's new leaders advocated greater decentralisation and autonomy for municipalities to better serve the population. In the case of water supply, the aim was to ensure the proper functioning of the infrastructure, the quality and quantity of water available, and an appropriate price for it, if necessary, through a takeover over the companies that managed the infrastructure and distribution.

However, the arguments in favour of municipalisation did not stem only from political-ideological convictions. The example of Lisbon demonstrates that it was used on several occasions throughout the nineteenth century as a form of pressure on the company that operated the city's water supply, in situations of conflict between the municipality and that company (Silva and Cardoso de Matos 2004). In other circumstances, illustrated by the case of Evora, although the municipalising option was not alien to republican ideological principles, the study of the process of establishing modern water supply infrastructures in the city demonstrates that the municipality did not override or hinder free initiative: it only had to perform functions in a field that did not seem of particular interest to private economic agents (Bernardo 2004). In the absence of interest from companies and in view of the growing technological and financial requirements associated with the installation and maintenance of these infrastructures, especially in projects that dealt with both water supply and sanitation, municipalities created municipal services within their administrative structure. This was an option aimed at technical and functional efficiency.

When an authoritarian regime was established in Portugal in the early 1930s, based on centralising principles at an administrative level and on economic interventionism, the municipalities lost their autonomy and became mere agents of the dictatorial state's guidelines. The structures of the local municipal services, and their technicians, served the regime's purposes by making the proposals of a technocratic modernisation inscribed in an authoritarian policy matrix stand out to the populations (Rosas 2000). It is not unlikely that the municipalisation option for water supply infrastructures in Spain, from the 1940s onwards, with the establishment of Francoism, was part of the same logic of appropriation of a technocratic discourse at the service of a political project of an authoritarian and centralising nature. The recognised "function of technology in nation building" (Saraiva 2007), in the historical circumstances referred to, would be at the service of *regime building*.

References

Amorim, A. A., and J. N. Pinto. 2001. *Porto d'Agoa: o abastecimento de água à cidade do Porto através dos tempos*. Porto: Serviços Municipalizados de Água e Saneamento.

Bernardo, M.A. 2004. "Elites, Acção Pública e Infra-estruturas: a construção da moderna rede de saneamento em Évora (1890–1933)." *Ler História* 46: 193–222.

Bernardo, M. A. 2013. *Sociedade e elites no concelho de Évora. Permanência e mudança (1890–1930)*. Lisboa: FCG/FCT.

Bernardo, M.A. 2019. "A questão da eletrificação urbana no debate político em Portugal: uma reflexão a partir dos debates parlamentares da primeira metade do século XX." In *La electricidad y la transformación de la vida urbana y social*, edited by H. Capel and M. Zaar, 566–80. Barcelona: Universidad de Barcelona/Geocrítica.

Brandão, J. M., and P. M. Callapez. 2017. *O abastecimento de água à Figueira da Foz em finais de oitocentos. Comodidade e modernidade*. Figueira da Foz: Ed CMFF.

Cardoso de Matos, A. 2019. "A ciência e a técnica ao serviço de cidades mais modernas e salubres." In *Contexto Programa Projeto: Arquitetura e Políticas Públicas de Habitação*, edited by R. J. G. Ramos, V. B. Pereira, M. Rocha, and S. D. Silva, 14–22. Porto: U. Porto Press.

Cliftona, J., P. Lanthier, and H. Schroter. 2011. "Regulating and Deregulating the Public Utilities 18302010." *Business History* 53 (5): 659–72. https://doi.org/10.1080/00076791.2011.599592.

Cordeiro, J. M. 2018. *História do Abastecimento de Água a Braga (1913–2013)*. Braga: AGERE – Empresa de Águas Efluentes e Resíduos de Braga.

Faria, R. de. 2015. "Urbanismo e municipalismo na Espanha: entre o estatuto municipal e a unión de municipios Españols na década de 1920." *Revista Brasileira. Estudos Urbanos e Regionais* 17(1): 105–22. https://doi.org/10.22296/2317-1529.2015v17n1p105.

Fernández-Paradas, M., and J. M. Matés-Barco. 2022. "Ingenieros y empresarios en las compañías de abastecimiento de agua en España (1840–1930)." *Aportes. Revista de Historia Contemporánea* 108, XXXVII: 75–111.

Fettah, S., and D. Bocquet (eds.). 2007. *Réseaux techniques et conflits de pouvoir: les dynamiques historiques des villes contemporaines.* Rome: École française de Rome.

Fougères, D. 1998. "Une approche sociotechnique pour l'étude historique des infrastructures et des services urbains: l'exemple du service d'eau à Montréal." In *Barcelona - Montréal. Desarrollo urbano comparado*, edited by H. Capel and P. A. Linteau, 202–3. Barcelona: Universidad de Barcelona.

Konvitz, J., M. Rose, and J. Tarr. 1990. "Technology and the City." *OAH Magazine of History* 5 (2): 32–7. https://doi.org/10.1093/maghis/5.2.32.

Lees, A., and H. L. Lees. 2013. "Europe: 1800–2000." In *The Oxford Handbook of Cities in World History*, edited by P. Clark, 464–82. Oxford: Oxford University Press. https://doi.org/10.1093/oxfordhb/9780199589531.013.0025.

Matés-Barco, J. M. 2004. "The Development of Water Supplies in Spain. 19th and 20th Centuries." In *Urban Growth on Two Continents in the 19th and 20th Centuries: Technology, Networks, Finance and Public Regulation. Introduction*, edited by A. Giuntini, P. Hertner, and G. Núñez, 165–77. Granada: Comares.

Matés Barco, J. M. 2021a. "Small, Medium, and Large Companies in the Supply of Water in Spain (1840–1940)." In *Entrepreneurship in Spain. A History*, edited by J. M. Matés-Barco and L. Caruana de las Cagigas, 82–98. New York-London: Routledge. https://doi.org/10.4324/9781003126973-7.

Matés Barco, J. M. 2021b. "Public Services in Spain: The Role of Water Supply Companies." In *Companies and Entrepreneurs in the History of Spain. Centuries Long Evolution in Business since the 15th Century*, edited by M. Vázquez-Fariñas, P. P. Ortúñez Goicolea, and M. Castro-Valdivia, 135–59. Palgrave Studies in Economic History. Cham: Palgrave Macmillan. https://doi.org/10.1007/978-3-030-61318-1_9.

Mendes, J. M. 2013. *Águas do Mondego e a sua história: tradição e inovação na captação e tratamento de água.* Coimbra: Águas do Mondego.

Milward, R. 2000. "The political economy of urban utilities." In *The Cambridge Urban History of Britain*, vol. 3, edited by M. Daunton. Cambridge: Cambridge University Press. https://doi.org/10.1017/CHOL9780521417075.012.

Milward, R. 2005. *Private and Public Enterprise in Europe. Energy, Telecommunications and Transport, 1830–1990.* Cambridge: Cambridge University Press. https://doi.org/10.1017/CBO9780511497124.

Montenegro, A. P. de M. 1903. *Inquérito de salubridade das povoações mais importantes de Portugal.* Lisboa: Imprensa Nacional.

Munck, B. de. 2017. "Re-Assembling Actor-Network Theory." *Urban History* 44: 111–22. https://doi.org/10.1017/S0963926816000298.

Newbery, D. M. 2002. "Regulating Unbundled Network Utilities." *The Economic and Social Review* 33 (1): 23–41.

Nunes, A. B., and N. Valério. 1996. "Lei da Reconstituição Económica." In *Dicionário de História do Estado Novo*, edited by F. Rosas, and J. M. Brandão de Brito, 512–3. Lisboa: Bertrand.

Nunes, A. B. 1996. "Portuguese Urban System: 1890–1991." In *Urban Dominance and Labour Market Differentiation of a European Capital City. Lisbon 1890–1990*, edited by P. T. Pereira, and M. E. Mata, 7–47. London: Kluwer Academic Publishers. https://doi.org/10.1007/978-94-011-5382-9_2.

Orduña Rebollo, E. 2006. "La Unión de Municipios Españoles: antecedentes y evolución del municipalismo asociativo." *Revista de Estudios de la Administración Local y Autonómica* 300–301: 331–60. https://doi.org/10.24965/reala.vi300-301.9300.

Pérez-Del-Hoyo, R., and M. E. Gutiérrez-Mozo. 2013. "Primeras políticas de vivienda en España y su influencia en la evolución de la tipología residencial: el caso de Benalúa (1883–1956)." *Revista INVI* 28 (78): 222–55. http://dx.doi.org/10.4067/S0718-83582013000200007.

Pinto, L. L. 1989. *Subsídios para a História do Abastecimento de Água à Região de Lisboa*. Lisboa: Imprensa Nacional-Casa da Moeda.

Pozas, L. J. 1961. *Estudios de Administración Local y General*. Madrid: IEAL.

Rodríguez Ocaña, E. 1994. "La salud pública en España en el contexto europeo, 1890–1925." *Revista de Sanidad e Higiene Pública* 68 (0): 11–27.

Rosas, F. 1986. *O estado novo nos anos trinta 1928–1938*. Lisboa: Estampa.

Rosas, F. 2000. *Salazarismo e fomento económico*. Lisboa: Editorial Notícias.

Saraiva, T., L. Schmidt, and J. Pato. 2014. "Lisbon Water regimes: Politics, Environment, Technology and Capital (1850–2010)." *Flux* 97/98: 60–79. https://doi.org/10.3917/flux.097.0060.

Saraiva, T. 2007. "Inventing the Technological Nation: The Example of Portugal (1851–1898)." *History and Technology* 23 (3): 263–73. https://doi.org/10.1080/07341510701300346.

Serra, J. B. 1988. "As reformas da administração local de 1872 a 1910." *Análise Social* XXIV (103–104): 1037–66.

Silva, A. F. 1997a. "A evolução da rede urbana portuguesa (1801–1940)." *Análise Social*, XXXII (143–144): 779–814.

Silva, A. F. 1997b. "Crescimento urbano, regulação e oportunidades empresarias: a construção residencial em Lisboa, 1860–1930." PhD diss., European University Institute, Florence.

Silva, A. F. 2020. "Thirsting for Efficiency: Technological and Transaction-Cost Explanations for the Municipalisation of Water Supplies." In *The History of Water*

Management in the Iberian Peninsula, Trends in the History of Science, edited by A. Duarte Rodrigues, and C. Toribio Marín, 89–110. Switzerland: Springer Nature. https://doi.org/10.1007/978-3-030-34061-2_5.

Silva, A. F., and M. L. Sousa. 2009. "In Search of the Urban Variable: Understanding the Roots of Urban Planning in Portugal". *Métropoles*, 6. Accessed 25 October 2022. http://journals.openedition.org/metropoles/4029.

Silva, A. F., and A. Cardoso de Matos. 2004. "The Networked City: Managing Power and Water Utilities in Portugal, 1850s–1920s." *Business and Economic History - On Line* 2: 1–45. Accessed 25 October 2022. http://www.thebhc.org/publications/BEHonline/2004/daSilvaMatos.pdf.

Sousa, F. de 1995. *História da estatística em Portugal*. Lisboa: Instituto Nacional de Estatística.

Tarr, J., and G. Dupuy (eds.). 1988. *Technology and the Rise of the Networked City in Europe and America*. Philadelphia: Palgrave.

Tostões, A. 1996. "Ministério das Obras Públicas." In *Dicionário de História do Estado Novo*, edited by F. Rosas, and J. M. Brandão de Brito, 585–98. Lisboa: Bertrand.

Veiga, T. R. 2003. "A população portuguesa no último século: permanências e mudanças." *Ler História* 45: 91–109.

Vleuten, E. van, R. Oldenziel, and M. Davids. 2017. *Engineering the Future, Understanding the Past: A Social History of Technology*. Amsterdam: Amsterdam University Press. https://doi.org/10.5117/9789462985407.

CHAPTER 4

Public Health, Sanitary Movement and Public Water Supply Services in the Nineteenth Century: The Cases of England and Spain

Nuria Magaldi

Abstract

This paper aims to analyse how the social hygiene movement was crucial to develop the first public mechanisms of water supply in Europe. In the nineteenth century, public health had mainly a preventive dimension: its main goal was to preserve society from epidemic diseases and provide it with better living conditions. To achieve this goal, the English sanitary movement (that spread throughout Europe, including Spain) firmly proposed to substitute traditional public fountains by a modern system of water supply. The final part of the paper, consequently, discusses how a modern public service of water supply was created and managed in many Municipalities in Spain, paying special attention to the Andalusian region.

Keywords

municipalities – public health – public water supply – sanitary movement

1 Introduction*

Today there is no doubt as to the direct relationship between people's health and the hygienic conditions of the places where they live. Among these conditions, the guaranteeing of an adequate provision and supply of drinking water, as well as an efficient sewage disposal system, occupy a prominent place. In this regard, it is well known that the modern system of urban water supply, which

* Research Project "Seguridad pública, actividad administrativa de limitación y derecho sancionador", PGC2018-093760-B-00 (MCIU/AEI/FEDER, UE) and "La Administración sancionadora de la Unión Europea", PID2022-138118NB, Spanish Ministry of Science.

 | DOI:10.1163/9789004679634_005

replaced, beginning in the second half of the nineteenth century, the *Ancien Regime*'s traditional system of public fountains, originated in Great Britain (Ramos and Rosado 2015, 1–4; Novo 2002, 295–6). In fact, it was in that country where, under the leadership of Edwin Chadwick,[1] the "Sanitary Movement" got underway during the first part of the nineteenth century, which would later lead to the development of modern running water supply systems and sewerage networks.

Several decades after it spread across Great Britain, the sanitary trend eventually reached our country. In spite of our late and haphazard industrialisation, the need for modern systems for the supplying and provisioning of drinking water also arose among us, as well as the discussion as to whether the management of this service should be placed in public hands or provided by private companies. This paper deals precisely with the sanitary movements in Great Britain and in our country, as well as their relationship with the development of modern public management systems for the supplying and provisioning of drinking water.

2 Sanitary Movement and Drinking Water Supply in the English Cities of the Nineteenth Century

2.1 *The Sanitary Movement and Its Link to Municipal Trading*

By the middle of the nineteenth century, the Industrial Revolution had turned many English cities into large industrial urban centres where large numbers of people having emigrated from the countryside were concentrated. This growth took place in an absolutely chaotic manner and without any public regulation whatsoever. Thus, houses were built with little separation between streets, and with scant ventilation, drawbacks aggravated by the absence of parks or green areas. Sewage systems did not exist, and there was no public rubbish and other waste collection service. Rather, only some private companies performed this task, which in no case served cities' impoverished, peripheral areas despite the fact that it was precisely these areas where the population from the countryside was agglomerated. As far as the water supply was concerned, when it existed, it

1 Regarded as the Father of Public Health, this British lawyer and economist had a short, but successful time in public office. In 1832 he was appointed Assistant Commissioner to the Poor Law Inquiry, a Commission established to make recommendations for reducing the cost of pauper programs in England. He retired from public service in 1854, while holding the position of Commissioner of the General Board of Health (Ringen 1979, 107–8).

was provided by private companies, almost always in a deficient and unhealthy manner (Ringen 1979, 113–4; Novo 2002, 286–7). As a result, living conditions in these cities deteriorated rapidly and health problems became widespread, including epidemics and cholera outbreaks.

There was a particularly serious problem regarding the use of cesspools, on the one hand, and sewage systems, on the other (Ramos 2014, 15). With regard to the latter, since they were designed only for river water, they were incapable of handling the solid waste which, with the huge shift of population from the countryside to the city, was constantly increasing, frequently clogging the system. As for the cesspools, they were not emptied regularly, nor were they designed to absorb the large volume of excrement generated in increasingly densely populated neighbourhoods, resulting in leaks and high levels of contamination with faecal matter. It was in this context that the lawyer and economist Edwin Chadwick published his influential *Sanitary Report* (1842), the landmark text in the English Sanitary Movement. Emphasising environmental conditions, he pointed to insalubrity – based on miasmatic theory – as the origin and cause of epidemic diseases such as typhus, typhoid fever, and cholera, and these, in turn, as one of the reasons why individuals often fell into destitution (Ramos and Rosado 2015, 2–3; Urteaga 1985–1986, 420–1). Contrary to the mainstream of his time, Chadwick believed that extreme poverty, in and of itself, was not the determining factor for these diseases since they did not really distinguish between the poor and the rich. If life expectancy among the low-income working classes was lower than among the upper classes, it was because they lived in overcrowded, damp settings, without adequate water supply and with serious deficiencies in the disposal of waste and refuse (Ramos 2014, 15–8).

In response to these circumstances, Chadwick endorsed intervention by public authorities to remedy unhealthy conditions through prevention, which would be achieved by improving air quality through the enhanced ventilation of living quarters, the development of a comprehensive sewage system, and the sufficient distribution of drinking water to homes (Ringen 1979, 116). Chadwick's "sanitary reform" was based, in accordance with his markedly utilitarian ideology, on the clear economic benefits of adequate disease prevention through the improvement of hygienic and sanitation conditions in cities. Accordingly, he emphasised the socioeconomic importance of equipping homes with direct running water and installing adequate sewerage and wastewater treatment systems, and how, in this way, the costs incurred by the weakening of the labour force, and the maintenance of widows and orphans, could be avoided. He also pointed out that urban wastewater, duly treated, could be reused to fertilise agricultural land, and that the loss of productive work time caused by long lines at public fountains could be avoided (Ramos 2014, 19–21; Ringen 1979, 114–5). Apart from this, worth mentioning is his firm commitment to public management of the provisioning and supply of drinking water in cities.

The ideas of Chadwick's Sanitary Movement coincided with the advent of the British "Municipal Trading" phenomenon, with which it is, undoubtedly, related (Magaldi 2010, 11–4). As already indicated, it was in Great Britain that the Industrial Revolution first took place, starting in the second half of the eighteenth century, with all the demographic, economic and social consequences that it entailed. This explains why that country was the first where the opportune and necessary circumstances converged for the emergence and development of the phenomenon of public industrialism; particularly, that of local public authorities, which was dubbed "Municipal Trading". This was a formula that consisted, essentially, in the cities themselves managing, directly and by themselves, the provisioning of certain services of a purely industrial nature, particularly the supply of drinking water. To this also contributed the peculiar configuration of the English governmental system, characterised by the weakness of its central authority, which erected, as a reaction, an extensive set of local government authorities that encompassed those functions that on the Continent were assigned to the State (Boverat 1912, 46; García 1955, 100–4).

What is certain, in this regard, is that the history of British municipal industrialism had its origins in the field of gas and water supply. With respect to the latter, there is also agreement that health and hygiene considerations were decisive in the public authorities' assumption of responsibility for their provisioning (Warren 1923, 13). In fact, Chadwick himself had already initiated the debate on the best way to organise the management of drinking water supply, stating in his *Sanitary Report* that the problems of supply and sanitation should be directly tackled by public authorities (Ramos and Rosado 2015, 5). His proposal clashed, however, with the situation that existed during the first third of the nineteenth century, dominated by the actions of private companies in a free water supply market, though there was growing unease among consumers and local managers.

In the following section we will discuss precisely how the provisioning and supply of drinking water evolved in British cities, in a process that began dominated by private companies operating in free competition and culminated with the assumption by local managers of all the major services that became essential in the "modern city" (Magaldi 2012, 27–35; Posada 1913).

2.2 *From (in)Effective Competition to Regulated Monopolies, and from Regulated Monopolies to Municipal Trading to Ensure Drinking Water Supply*

Prior to the nineteenth century, some English cities did manage some services considered essential for municipalities, such as markets and slaughterhouses, and had even undertaken supplying their populations with water. However, it cannot be said that this was a strictly corporate activity. Thus, it was at the beginning of the nineteenth century when we must situate the first actions that could be classified as "municipal industrialism", coinciding with the development of large cities and agglomerations, in which the provisioning of water, electricity and transport became necessary, with these activities taking on an eminently industrial nature (Warren 1913, 13).

During this first period the work of lighting, watching, and paving commissioners was key, these being city officials in charge of overseeing the provisioning of lighting, sewage, and water supply services. The truth is that some of these positions were really ahead of their time, as they clearly reflected an awareness of a responsibility for the provisioning of certain services, particularly the supply of water and gas (Warren 1923, 14). Thus, for example, the Commissioner of Leeds began to supply water in 1790 and the Commissioners of Manchester were particularly active in relation to the production and supply of gas for municipal use.

Though these commissioners were, undoubtedly, pioneers and ahead of their time, they were an exception in a scenario dominated by the manoeuvring and activities of private companies. Practically the entire first third of the nineteenth century in England was characterised by a growing tension between consumers (due to poor service), private companies (which acted freely in the various markets) and local managers (who saw how habitability and public hygiene conditions in the cities under their responsibility were drastically worsening).

To correctly understand this situation of discord, it is necessary to bear in mind the specific characteristics of a modern drinking water supply service. Significant initial investments were required, inasmuch as it was necessary to install, beforehand, pipes, wires and distribution pipelines for its provision, but the companies interested in doing so needed authorisation from the public authorities, since these pipelines or conduits normally crossed public land (Boverat 1912, 64–5; Shaw 1889, 220). Thus, although there was a first phase in which private acts were granted to all the water companies that requested them, it was soon realised that this was extremely wasteful for the businesses involved, and that it did not favour consumers either. The private companies ended up building the same facilities in the same area, or on the same street, which meant two or more very costly initial capital investments. Then, since each company passed on the installation costs to consumers in their prices, they ended up footing the bill for multiple installations and their maintenance (Warren 1923, 16).

These services, moreover, followed the theory of decreasing economies of scale, therefore tending towards monopolisation, as whoever first obtained authorisation enjoyed an advantageous position with respect to other possible competitors seeking to enter the market at a later date. This was a situation verging on monopoly and it developed quite rapidly as the various competing companies joined forces with each other or reached agreements to divide up their different areas of influence and supply. In fact, in 1860 London's thirteen water supply companies presented a bill in Parliament requesting, basically, the organisation of supply by districts (Warren 1923, 19).

This period during which monopolies and oligopolies predominated, without any limitations on the part of the public authorities, generated serious disadvantages, especially for consumers, who had to endure the monopolistic prices imposed by the concessionaires and, all too often, deficient service in terms of quantity, quality, or both. Between 1821 and 1840, numerous complaints were submitted to the Select Committees or the Royal Commissions by consumers, acting through the local authorities, most of which were well founded, and these bodies systematically proposed the need to establish a regulatory framework, but for many years Parliament failed to legislate on the matter (Warren 1923, 20; Albi 1960, 71–2).

In 1844, all this resulted in a very harsh report issued by the Royal Commission of The Health of Towns, which evaluated in very negative terms the supply of drinking water in the main English cities. This Commission had been created after the publication of the *Sanitary Report* to analyse what it stated and to evaluate its recommendations. In particular, the report placed special emphasis on the ruinous consequences that competition between companies had spawned in the first place. It also pointed out the harmful consequences for service users of unregulated monopolies or oligopolies. Consequently, it recommended the establishment of conditions and limits on concessionary companies; for example, the establishment of limitations on the dividends they could obtain, and minimum quality standards (Warren 1923, 20–1; Boverat 1912, 68). The following year the Royal Commission of The Health of Towns published a new report along similar lines, which, together with the previous one, would pave the way for the Public Health Act of 1848 – a piece of legislation that, to a large extent, was based on the positions articulated by Chadwick – (Ringen 1979, 114; Ramos 2014, 22–4). It also contained the creation of a General Board of Health which, as an autonomous state entity, would intervene in local health matters when a certain mortality rate was exceeded in a city.

Given this situation, it did not take long for a general movement to emerge in favour of the direct municipal provisioning of the main services, starting with the supply of drinking water, such that many cities began to acquire the existing facilities, through amicable agreements with private companies, or by submitting the matter to arbitration, as was the case of Sheffield in 1830 (Willis 2001, 103), Manchester in 1847 (Boverat 1912, 94–8), Glasgow in 1855 (Shaw 1889, 221–2; Pedregal 1930, 34), and Birmingham in 1876 (Elorrieta 1919, 10). In other cases, if there was no prior concession, the municipalities opted to build their own facilities directly (Bath, Coventry, Halifax, Plymouth). Although the laws did not provide special terms for the transfer of water companies from the private to the public sector, the policy of Parliament was not to refuse the corresponding authorisation from any local authority seeking to obtain control of this service, on the basis that it was a service that fell within the duties of the municipality (Boverat 1912, 82).

Of special relevance was the very extensive Public Health Act, approved in 1875, which, consolidating some previous regulations (the Water Act in 1845 and the Public Health Act in 1848) laid the foundations for the development of local health systems. This regulation empowered municipal governments to act as health authorities, appointing local "medical officers of health" and "inspectors of nuisances". It also assumed responsibility for control of the sewage system, the hygiene of buildings, and medical care for the needy population, as well as for cemeteries, public gardens, markets and slaughterhouses, workers' homes, etc. It also granted them, in general, the power to build and operate water and gas installations (Dogliani 1988, 223; Warren 1923, 21; Boverat 1912, 51–2).

Thus, in 1902, a total of 193 English and Welsh municipalities had their own water companies (including Bath, Coventry, Halifax, Hastings, Huddersfield, Hull, Liverpool, Oxford, Plymouth, Southampton, and Worcester), and there were 1905 municipal water companies across the entire United Kingdom. London, one of the last municipalities to take over the service, also did so in 1902, with the creation of the Metropolitan Water Board for the municipal water supply (Boverat 1912, 81–3; Gascón 1904, 139–51).

3 The Reception of the Sanitary Movement in Spain

Spain was not unaware of the sanitation ideas that spread throughout Europe in the nineteenth century, although it was slow to embrace and implement the public health ideas coming from other European countries, particularly Great Britain and France. In that regard, the work of a forerunner in the field of sanitation in Spain, Mateo Seoane Sobral, and that of his two main disciples, Pedro Monlau y Roca and Francisco Méndez Álvaro, exhibited a marked influence by the Sanitary Movement. In 1837 Mateo Seoane published his work *Consideraciones generales de estadística médica* (*General Considerations of Medical Statistics*), deeply influenced by the spirit of the British movement, which he had imbibed during the decade he had spent in London. With this work he became one of the leading scientific figures in our country with regard to the organisation of the medical profession, medical teaching, and, above all, the provisioning of public hygiene (Alcaide 1999), transmitting his knowledge of the British movement to his disciples Monlau and Méndez Álvaro. The latter, in fact, demonstrated in his works to be well versed in the British public health project, making numerous references to Chadwick himself. Monlau, meanwhile, apparently was more familiar with the sanitary movement in France, due to his stay in Paris from 1837 to 1839, but it should not be forgotten that there was also a major process of mutual influence between the British and French public health movements, especially from the 1840s onwards, when Chadwick's propositions became popular among French treatise writers (Alcaide 2017, 55–6).

Spanish sanitation advocates, in any case, took as their starting point the basic propositions of the British school. Like Chadwick, they believed in the greater impact of disease and mortality on the urban/industrial classes, stating that, although industrial work itself was not a direct cause of disease, the conditions of poor ventilation and overcrowding in which it was carried out were. Consequently, they focused their attention on improving the urban-industrial environment, pointing to the need to carry out specific actions to improve sanitation in workshops, factories, and homes; and to prioritise the supplying of abundant water, the creation of sewage networks, and rubbish collection (Ramos 2014, 26–9). Like Chadwick, to explain disease they invoked miasmatic theories, but they did not limit their explanations for the origin of diseases and epidemics to insalubrity, like Chadwick, but also considered poverty a relevant factor. They thus positioned themselves halfway between the UK's Sanitary Movement and the French sanitation advocates, who had drawn attention to the social dimension of disease, linking infectious diseases such as tuberculosis with the lack of clean air, overcrowding and malnutrition that accompanied urban poverty. Here, Monlau's particular vision stands out, as he went so far as to propose the decentralisation of factories and the ruralisation of industry, flatly condemning large urban-industrial agglomerations and extolling the supposed moral and hygienic superiority of peasants (Rodríguez and Menéndez 2005, 59; Ramos 2014, 29–31).

In spite of what has been said, the truth is that the Spanish sanitary movement in the mid-nineteenth century lacked an empirical approach based on its own direct data and testimony. Rather, it followed what was happening in other countries (Rodríguez and Menéndez 2005, 59). Its considerations on urban sanitation infrastructures were limited to a more theoretical and generic level, so its institutional scope was much more limited than in the British case, and its concrete achievements in the short term were much less tangible and significant. This was probably related to the fact that, in Spain, industrial and urban development, and the consequent socioeconomic transformation, were late to come and concentrated in only a few regions. Thus, the context in both countries was radically different, such that the development of modern public health, linked to cities, could not, in any case, take on the same dimensions in our country as in Great Britain. This also helps to explain why the nineteenth century Spanish sanitary movement, despite emphasising the preventive dimension, never clearly clarified what the role of the public authorities in the construction of sanitation infrastructures should be. Certainly, the scientific work of Seoane, Monlau and Méndez Álvaro contributed decisively to the development of the liberal sanitation system, but this would not incorporate, until much later, the debate on whether water supply infrastructures should be provided by local public authorities, as was the case in Great Britain. We will discuss this issue in greater detail in the following section.

4 Drinking Water Supply Services in Spain (1850–1930): Special Reference to Andalusia

4.1 *Spanish Backwardness in the Public Management of Drinking Water Supply Services*

Unlike what happened in Great Britain, in Spain it is not possible to establish solid, direct links between the ideas propounded by the sanitary movement and the takeover of public services by local authorities. In Spain the supplying of drinking water fell within the municipalities' scope of activity, with nineteenth century regulations granting them authority in the matter. The first legislative manifestation of the local nature of water supply and sanitation services can be found in the Royal Decree for the Provisional Arrangement of the Local Governments of the Kingdom, dated 23 July 1835, whose article 48.4, third paragraph, described as "powers to local governments" seeing to the "health, cleanliness and adornment" of towns and cities, which would include sanitation and the evacuation of sewage. Likewise, article 48.6 established as powers to local governments "securing the best possible supply of potable and abundant water for public use". However, barely a decade later, the Law of 8 January 1845, on the Organisation and Powers of Local Governments, did not include water supply and sanitation among municipal competencies, which has been clearly attributed to the "statocentric" conception of administrative law prevailing during that period (Ezquerra 2007, 287–8).

In any case, the Municipal Law of 2 October 1877, in its article 72, in a change of direction with respect to the previous regulation of 1845, was clear in charging local governments with "the exclusive establishment and creation of those municipal services related to the arrangement and ornamentation of public streets, the comfort and hygiene of neighbourhoods, and the promotion of the material and moral interests and safety of their persons and property", expressly including among these services "water supply" and the "people's cleanliness, hygiene and health". This municipal competence was reinforced by art. 137 2º of the same legal body, which considered the possibility of authorising local governments to levy taxes for, among other services, the "use and supply of water for private use" and the "sewerage system".

By the beginning of the twentieth century, the Health Directive of 12 January 1904 stated that municipal hygiene included "the supplying of water and the monitoring of its purity in tanks, pipes and springs", as well as "the evaluation of sewage and wastewater". Meanwhile, the Municipal Sanitation Regulation of 9 February 1925 established as a primary obligation of municipalities to provide their populations with drinking water sufficient for the necessities of life, as well as to establish and maintain a system for the evaluation of wastewater (Ezquerra 2007, 295–6).

The turning point came, undoubtedly, with the Municipal Statute under Calvo Sotelo, approved in 1924, which established a new system for water supply, sewage disposal and treatment services that was later widely adopted by subsequent local regulations (Ezquerra 2007, 290). Its article 150 declared the exclusive responsibility of local governments in "the administration and management of their peculiar interests" and, more specifically, "the supplying of water and the processing of wastewater" (section 9) and "sewers, the draining of lagoons or swamps within the municipal boundaries, and any other health and hygiene services" (section 10). Likewise, and specifically in relation to the works required for the provisioning of the services, article 180.2 classified as an "exclusive municipal competence" the design, construction, and approval of the plans for works designed for "the supplying of water, sewerage and wastewater treatment services for the population centres located in the corresponding municipal districts, or to improve the quality of said services". Particularly relevant was the fact that, in addition to assigning local governments authority over matters of domestic water supply and disposal, both services were classified as compulsory for all municipalities, regardless of their populations (article 202). Finally, article 169 of the Statute expressly allowed the municipalities to take on the direct administration and operation of these services.

Despite all this, our country did not manage to follow in the footsteps of the British, as very few towns and cities directly took over responsibility for the supplying of water to individuals. This was due, at least in part, to the general weakness in Spain of the phenomenon equivalent to Municipal Trading, which in our country was referred to as the "municipalisation of services" (Magaldi 2012), with this shortcoming being evident in the case of the supplying of drinking water. In Spain, it is not possible to speak of a reasonably consolidated public water sector in the period between 1850 and 1930. The truth is that, for many reasons, Spanish local governments of that period proved incapable of exercising their municipal powers, leaving the issue of water supply in the hands of private companies, through the granting of concessions. As we shall see below, in Spain, during the period mentioned above, there was a great expansion of private companies (often foreign), which sought to take on responsibility for managing and solving the country's water supply problems (Matés-Barco 1998, 197–201; 2000, 38; Mirás 2003, 474; Núñez 1996, 409). To this must be added the country's late industrial, economic, and urban development, which, until the period from 1900–1930, was not significant or widespread enough for the need for drinking water supply to constitute a major problem (Membiela 1950, 127–8). It was not until 1925, after the approval of the Statute under Calvo Sotelo, that local entities began, in a widespread manner, to take charge of the supplying of the service, coinciding in many cases with the end of the concessions or, directly, their takeover by local governments.

Generally speaking, in those municipalities in which municipal needs were suitably met by private companies, local governments did not directly take on responsibility for the provisioning of the service (they did not municipalise it), aware of the economic burden that the expropriation of the private companies entailed. Only when private companies did not perform satisfactorily were attempts made to intervene and secure the service (Membiela 1950, 127–8). Thus, the progressive public assumption of management of the service (i.e., its municipalisation) took place, in our country, in the period between 1940 and 1970, when the system's internal contradictions (half public, half private) led to the gradual extinction of the private companies and their replacement by public initiatives (Matés-Barco 1998, 127; 2000, 52).

4.2 *Drinking Water Supply Services in the Main Spanish Cities*

In the period studied here, we can indicate that the water supply service was not provided directly by the city governments of Alicante, Badajoz, Barcelona, Burgos, Badalona, Cáceres, Ceuta, A Coruña and Pamplona (Sociedad de

Conducción de Aguas de Arteta 1935), among others, while Oviedo (Matés-Barco 2008, 278), Salamanca, Vitoria, Bilbao (Fernández 2001, 39–41), Castellón, and Lugo (Matés-Barco 2008, 277) municipalised it, providing high-quality service, while Elche and Melilla municipalised it, but offered deficient service.

Zaragoza began to intervene in water provisioning in 1876, when it started to collect water from the "Imperial Canal", basically used for public fountains. The city's rapid demographic development increased the need for the service, which had to be expanded from 1907 onwards. In any case, the service was provided, from the beginning, by the city government, without mediation by any concessionary company (Membiela 1950, 138–42).

Palma de Mallorca's growth at the beginning of the twentieth century generated a serious water shortage, which, in turn, coincided with the city's demolition of its walls and expansion. In 1912 the city government legally acquired the spring that provided water, and in 1927 it proceeded to carry out a project of pressurised canalisation for certain areas, expanding it to the whole city in 1932. Therefore, the water supply in Palma was carried out, from very early on, by the municipal corporation, without intermediation by any concessionaire company (Enseñat 1920, 54–80).

In the city of Las Palmas de Gran Canaria the service was initially operated by a private company. In 1911, population growth experienced led to the assignment of the construction of the works and the subsequent management of the supply service to the *City of Las Palmas Water and Company Limited*, which initially made important advances in its quality. Over time, relations between the municipal corporation and the company deteriorated, and in 1930 the concession was rescinded, with a court decision dictating that the city was to pay the company 20 million pesetas. After years of lawsuits and disputes, the municipal corporation negotiated a more advantageous agreement, and ended up becoming the owner of the installations and the manager of the service (Membiela 1950, 145–9).

The work to bring water to the city of San Sebastián began in 1889, drawing on the waters of the Añarbe River, a tributary of the Urumea. The first problem arose when the river was found to be contaminated after an alarming typhus epidemic in the summer of 1902. It was, then, necessary to take drastic measures to improve the sanitation of the waters, and one of the solutions suggested was to obtain water from the springs on Mount Ariscutza. Thus, in 1903 the corporation initiated a process to exploit the spring waters there, giving rise to controversies and disputes that would not be resolved until the city's full acquisition of the land on 27 January 1919 (Larrinaga 2008, 87–90).

The first source of drinking water in Albacete was inaugurated in 1862, and was managed by a private company that was, at the same time, the concessionaire of the railroad (Membiela 1950, 150–70). The supply was deficient, flawed, and rudimentary, until the contract was terminated in 1875. During this same period, the city government carried out various projects through direct administration that bolstered its water catchment capacity, obtaining it from nearby springs that, via a Royal Order of 10 October 1901, were recognised as municipal property. In 1903, the service was transferred to the concessionaire *Aguas Potables de Albacete,* which was granted a 90-year concession, with all the works and installations reverting to the City Council at the end of this period. The new company combined the interest and spirit of a private company with respect for the public interest, carrying out the necessary works so that the service could be effectively launched in 1905. It was later improved, with moderate rates being applied in an attempt to ensure the service's affordability. In spite of all this, public opinion favoured municipalisation by buying out individuals' shares in the company, which was ultimately done at the end of 1916, with the service and its installations passing into the hands of the city government.

The situation in Spain's capital city was peculiar, as its drinking water supply service was provided by a public, state body, and only in some areas by private concessionaire companies. It should be noted that in 1851 the Government of Bravo Murillo had approved a Royal Decree providing that the Spanish state would proceed to directly carry out the work necessary to supply Madrid with healthy water, by means of a canal drawing from the Lozoya River, to be dubbed the Canal de Isabel II, which in 1867 became an ordinary service of the State. Over time, consumption increased, which complicated the service's technical administration. Thus, at the beginning of the twentieth century the Canal was granted greater autonomy, on the basis of a special Board of Directors charged with its management, immediately and directly attached and answerable to the Ministry of Public Works (Suárez 1927, 7). The first proposals for municipalisation were not put forward until the approval of the Municipal Statute of 1924, in an attempt to resolve the capital's worrisome water supply situation. In fact, the water conduction channel in 1925 was still the same as in 1858, but the population of Madrid had grown considerably, so there was a risk that an incident could leave part of the population without water (Zurano 1925, 3). It was in the context of this tense and distressing situation that Madrid's mayor proposed proceeding with municipalisation, and even initiated the process, but was unable to complete it (Martínez and González 1926, 10–4; Suárez 1927, 14–6).

4.3 *The Situation in Andalusia*

As far as the specific situation in Andalusia is concerned, and starting with larger municipalities that were not, however, provincial capitals, it should be noted that Linares municipalised the service in 1882. Algeciras did not provide it directly, but rather assigned it to an important private drinking water company, the *Andalusia Water Company Limited, S. A.*, incorporated in 1912. The company's limited profitability and the difficulties it went through explain why in 1916, it transferred its rights to supply the port and the Bay of Algeciras to the *Sociedad de la Bahía de Algeciras* (Matés-Barco 1997, 110). In Jerez de la Frontera, where the company *Abastecimiento de Aguas Potables de Jerez de la Frontera S. A.* was constituted in 1868, exclusively for the supplying of drinking water to that city, the service was not municipalised either, but the city did own 60% of the operating company's shares (Matés-Barco 1997, 110; 1998, 81–2 and 97–8).

The service was municipalised and provided quality service in the provincial capitals of Almeria, Huelva and Jaen. In Cordoba, the service was originally provided by a private company, *Empresa de Aguas Potables S. A.*, incorporated in 1891. This, however, did not cover the whole city, so it had to rely on other systems as well. Despite the company's able management, it could not overcome the difficult economic situation of the late 30s, and the service was eventually municipalised, with the *Servicio Municipal de Aguas Potables de Córdoba* taking over (Membiela 1950, 135–8; Matés-Barco 1997, 122–3).

At the beginning of the century, the city of Seville had granted all the major services – including gas and electric lighting, transport, sewerage and drinking water supply – to private companies (Ayala 1910). As regards the latter, the service was entrusted to the concessionaire *The Seville Water Works C*[o] *Ld*, comprised of foreign capital and incorporated in London, by means of a concession granted in 1882, to last for 99 years. This company specialised in the manufacture of cast iron pipes and was particularly interested in supplying the city in order to consolidate the marketing and distribution of its main product (Matés-Barco 1997, 105). The company provided the service while meeting the established commitments, undertaking works to build new springs, and providing purified water of good quality. However, the springs originally used were gradually depleted, and it was necessary to resort to complementary catchment through supplementary contracts with the city government. The truth is that the situation described above did not improve, so after the First World War the city government decided to directly undertake the works for a second distribution network, which was not completed until 1926 and was managed by the city. In this way, Seville was served by a dual network: the one carrying the water of the concessionaire company, and the water managed by the city. In any case, as of the middle of the 1920s there were many voices calling for the municipalisation of the service, which was not economically feasible (Matés-Barco 1997, 105–9).

As for the city of Cadiz (León de Carranza 1959; Matés-Barco 1997, 112–22), the stagnation of water in the bay made it necessary to bring in water from outside the city starting in the early eighteenth century, with the high costs that this entailed. After multiple failed projects throughout the first half of the nineteenth century, an agreement was finally reached between the city and a private party, Matías del Cacho, with whom a contract was signed in February 1868 that granted this businessman the water supply concession. The term of the concession was set at 99 years, after which del Cacho would transfer all his property rights to the city, without any kind of reservation or any right to any compensation whatsoever. The concession contract also indicated the prices of water for public consumption, and the obligation to cede a certain amount of water to the city free of charge. However, the difficulties involved far exceeded the concessionaire's technical and investment expertise, and he ended up ceding his rights to the English company *The Cadiz Water Works Company Ltd.* On 10 November 1871, the deed of cession was signed with the British company, which in March 1872 commenced the construction of a pipeline to bring water from the Sidueña Valley. There is no precise information on the costs involved or the investments made by the company, but it it seems that it suffered problems and difficulties similar to those faced by del Cacho. It could be that, once again, the contract signed with the company called for a volume of water the company could not deliver, such that it ended up declaring bankruptcy, and in December of 1876 the city declared the concession null and void. Shortly afterwards, by means of a public auction, a French company acquired the English company's assets. After the death of Cadiz's illustrious Diego Montañés, the city, taking advantage of his donation to it, acquired in 1882 the land, works and materials for its water supply. For some months it was the municipality itself that administrated the service, but budgetary deficiencies and the scarce resources had made it impossible to undertake a series of works that were considered essential to improve the supply. Faced with these difficulties, in 1884 the municipal corporation decided to assign the service via a public auction, awarding it to José Hernández Quintero on 30 June 1885. A few days later, the concessionaire transferred his rights to the *Sociedad de Aguas Potables de Cádiz*, with the city's approval. The new company undertook the works quickly, and the first water reached the city in August of 1887, but it continued to be insufficient in quantity and of poor quality. It was not until 1899 that the city took up the matter again, its disagreements with the supplying company becoming more and more serious, while the quality of the water continued to leave much to be desired, and the shortage had become chronic.

5 Conclusions

The protection of public health during the nineteenth century and the first third of the twentieth was, fundamentally, a collective effort. The aim was to prevent or, at least, mitigate sanitation deficiencies that could lead to infections and epidemics seriously jeopardising people's health, and even their lives. It is in this context that the sanitary movement emerged and spread from Great Britain (and France) to our country.

Authors rightly emphasised the inseparable link between public health and an adequate drinking water supply system that could guarantee suitably hygienic conditions. At the same time, local managers, especially in large urban centres, realised that the best way to ensure sanitation was through the direct provisioning of the water supply by the local authorities themselves. Thus, public drinking water supply services appeared, which became crucial elements in improving public health in the industrial urban centres that proliferated across much of the Continent, especially as of the second half of the nineteenth century.

References

Ayala, J. 1910. *Estudio de los servicios públicos que convendría municipalizar en Sevilla.* Sevilla: Ayuntamiento de Sevilla.

Albi, F. 1960. *Tratado de los modos de gestión de las corporaciones locales.* Madrid: Aguilar.

Alcaide González, R. 1999. "La introducción y el desarrollo del higienismo en España durante el siglo XIX: precursores, continuadores y marco legal de un proyecto científico y social." *Scripta Nova: Revista electrónica de geografía y ciencias sociales* 3: 32–54.

Alcaide González, R. 2017. "El higienismo y su implicación en el desarrollo del marco legal de los transportes públicos en España (1848–1936)." *TST, Transportes, Servicios y Telecomunicaciones* 33: 52–69.

Boverat, R. 1912. *Le socialisme municipal en Anglaterre et ses résultats financiers.* Paris: Arthur Rousseau.

Dogliani, P. 1988. "Il dibattito sulla municipalizzazione in Europa dall'inizio del novecento alla prima guerra mondiale." In *La municipalizzazione in area padana. Storie ed esperienze a confronto*, edited by A. Berselli, F. della Peruta, and A. Varni, 220–56. Milan: Franco Angeli.

Elorrieta y Artaza, T. 1919. *La municipalización de servicios en el Condado de Londres.* Conference on 13 March 1919. Madrid: Imprenta municipal.

Enseñat Alemany, J. 1920. *El Ayuntamiento de Palma y la municipalización*. Palma de Mallorca: Tipografía de Armengual y Muntaner.

Ezquerra Huerva, A. 2007. *El régimen jurídico de las obras hidráulicas*. Murcia: Fundación Instituto Euromediterráneo.

Fernández, A. 2001. "L'impossible socialisme municipal en Espagne." In *Der Munizipalsozialismus in Europa/ Le socialisme municipal en Europe*, edited by U. Kühl, 31–47. Munich: Parischen Historische Studien, Oldenbourg Verlag.

García de Enterría, E. 1955. "La actividad industrial y mercantil de los municipios." *Revista de Administración Pública* 17: 87–138.

Gascón y Marín, J. 1904. *Municipalización de servicios públicos*. Madrid: Librería General Victoriano Suárez.

Larrinaga, C. 2008. "Modernización y servicios urbanos en San Sebastián en el primer tercio del siglo XX." In *La modernización económica de los ayuntamientos: servicios públicos, finanzas y gobiernos municipales*, edited by L. González Ruiz, and J. M. Matés-Barco, 82–116. Jaen: Universidad de Jaén.

Magaldi, N. 2010. "Los orígenes de la municipalización de servicios. El industrialismo público inglés (*Municipal Trading*) y la Sociedad Fabiana." *Revista de Estudios de la Administración local y autonómica* 313–314: 11–54. https://doi.org/10.24965/reala.vi313-314.9921

Magaldi, N. 2012. *Los orígenes de la municipalización de servicios en España*. Madrid: Instituto Nacional de Administración Pública.

Martínez Ángel, M., and P. González Quijano. 1926. *Proyecto de municipalización del canal de Isabel II*. Report by the representatives of the Chamber of Industry and the Institute of Civil Engineers on the Commission appointed by the Madrid City Council.

Matés-Barco, J. M. 1997. "Las sociedades anónimas de abastecimiento de agua en Andalucía: una primera aproximación." *Boletín del Instituto de Estudios Giennenses* 167: 103–30.

Matés-Barco, J. M. 1998. *Cambio institucional y servicios municipales. Una historia del servicio público de abastecimiento de agua*. Granada: Comares.

Matés-Barco, J. M. 2000. "La conquista del agua: importancia urbana y económica." *Boletín del Instituto de Estudios Giennenses* 174: 29–55.

Matés-Barco, J. M. 2008. "Las empresas de abastecimiento de agua en la España del Norte (1840–1970)." In *La modernización económica de los Ayuntamientos: servicios públicos, finanzas y gobiernos municipales*, edited by J. M. Matés-Barco, and L. González Ruiz, 269–91. Jaen: Universidad de Jaén.

Membiela Guitian, A. 1950. *La municipalización de servicios públicos*. Madrid: Instituto de Estudios de la Administración Local.

Mirás Araujo, J. 2003. "La regulación del abastecimiento de agua en la ciudad de A Coruña. De la gestión privada a la municipalización." SEMATA, *Ciencias Sociales e Humanidades* 15: 473–86.

Novo López, P. 2002. "El agua potable a domicilio y los residuos urbanos bailan un complejo pas a deux (1850–1939): una aproximación metodológica e historiográfica." *Historia Contemporánea* 24: 281–319.

Núñez Romero-Balmas, G. 1996. "Servicios urbanos colectivos en España durante la segunda industrialización: entre la empresa privada y la gestión pública." In *La empresa en la historia de España*, edited by F. Comín and P. Martín Aceña, 399–419. Madrid: Civitas.

Pedregal y Fernández, M. 1930. *Municipalización comparada*. Madrid: Tipografía de Archivos.

Posada, A. 1979. "La ciudad moderna" [1913]. In *Escritos municipalistas de la vida local*, edited by A. Posada, 317–409. Madrid: Instituto de Administración Pública.

Ramos Gorostiza, J. L. 2014. "Edwin Chadwick, el movimiento británico de salud pública y el higienismo español." *Revista de Historia Industrial* 55: 11–38.

Ramos Gorostiza, J. L., and A. Rosado Cubero. 2015. "Ideas económicas en torno al servicio de abastecimiento urbano de agua en la Gran Bretaña del siglo XIX." *Investigaciones de Historia Económica* 11: 1–9. https://doi.org/10.1016/j.ihe.2013.09.001.

Ringen, K. 1979. "Edwin Chadwick, the Market Ideology and Sanitary Reform: On the Nature of the 19th Century Public Health Movement." *International Journal of Health Services* 2 (1): 107–20. https://doi.org/10.2190/LR4G-X2NK-9363-F1EC.

Rodríguez, E. and A. Menéndez. 2005. "Salud, trabajo y medicina en la España del siglo XIX. La higiene industrial en el contexto antiintervencionista." *Archivo de Prevención de Riesgos Laborales* 8 (2): 58–63.

Shaw, A. 1889. "The municipal government in Great Britain." *Political Science Quarterly* 4 (2): 197–229. https://doi.org/10.2307/2139337.

Sociedad de Conducción de Aguas de Arteta. 1935. *Sociedad anónima de conducción de aguas de Arteta: antecedentes imprescindibles y réplica a la memoria de la Comisión especial para el estudio de la municipalización del servicio de aguas*. Pamplona: Sociedad de Conducción de Aguas Arteta.

Suárez Inclán, H. 1927. *Municipalización y abastecimiento de agua en Madrid*. Madrid: Suc Rivadeneyra.

Urteaga, L. 1985–1986. "Higenismo y ambientalismo en la medicina decimonónica." *Dynamis: Acta Hispanica ad Medicinae Scientiarumque Historiam Ilustrandam* 5–6: 417–25.

Warren, J. 1923. *Municipal Trading*. London: The labour publishing company limited.

Willis, T. 2001. "Contributing to a Real Socialist Commonwealth: Municipal Socialism and Health Care in Sheffield (1918–1930)." In *Der Munizipalsozialismus in Europa/Le socialisme municipal en Europe*, edited by U. Kühl, 101–17. Munich: Parischen Historische Studien, Oldenbourg Verlag.

Zurano Muñoz, E. 1925. *El canal de Isabel II; peligros y gravámenes que supone para Madrid la municipalización* (lecture given at the Círculo de la Unión Mercantil e Industrial de Madrid, 30 November 1925). Madrid: Imprenta Juan Pueyo.

CHAPTER 5

Water Supply in the City of Cadiz in the Nineteenth Century: *The Cadiz Water Works Company Limited*

María Vázquez-Fariñas

Abstract

Water supply has been one of the most important services in Spain since the mid-nineteenth century. Therefore, this chapter presents an analysis of the public water supply service in the city of Cadiz (Andalusia, southern Spain) during the nineteenth century. Specifically, it studies the water supply in the city, the nature of the management and supply of this resource and the main characteristics of the supply. In addition, this work shows the group of private companies that emerged in the city of Cadiz and contributed to the development of the sector. In particular, the study analyses the role played by the London-based company *The Cadiz Water Works Company Limited* in providing the service for the management and supply of drinking water to the city of Cadiz. So, this chapter seeks to address the principal characteristics of this company and the sector, and the most relevant problems prevailing to determine how the water supply and management service was developed in the city.

Keywords

Cadiz – English investment – private company – *The Cadiz Water Works Company Limited* – water management

1 Introduction

The objective of this study is to analyse the role played by the London-based company *The Cadiz Water Works Company Limited* in providing the service for the management and supply of drinking water to the city of Cadiz (Andalusia, southern Spain) at the end of the nineteenth century.

 | DOI:10.1163/9789004679634_006

The provision of this public service changed considerably in Spain throughout the nineteenth century in terms of technology, forms of management and sanitation. A modern drinking water system was installed, characterised by the establishment of domiciliary networks with a permanent and continuous supply and the application of a tariff regime (Matés-Barco 2017, 37).

In the city of Cadiz, similarly to other large cities of the province, such as Jerez de la Frontera or Algeciras, there were companies that had been operating in the water supply sector for some years, which played a fundamental role in the provision of the service. Therefore, both private companies and the city council were important actors in the development of this sector as, although the service was initially established thanks to the private initiative, from the end of the nineteenth century a gradual municipalisation process began, until the management of the service was completely under the control of the municipality. One of the most relevant companies was *The Cadiz Water Works Company Limited*. This study seeks to address the principal characteristics of this company and the sector and the most relevant problems prevailing at the end of the nineteenth century so as to determine how the service of supplying and managing water in the city developed.

Many primary and secondary sources have been used to conduct this study. First, as a secondary source, the existing bibliography on the subject matter has been analysed. In this respect, fundamental studies include those by Matés-Barco (1997, 2014, 2017, 2020, 2021) on the water supply service in Spain and Andalusia; the study by Pérez, Román, Villatoro, and Arévalo (2013), addressing the supply of water in the Bay of Cadiz from the eighteenth century; the study by Pérez and Román (2015) on the conflicts over water resources in the Bay of Cadiz and the study by Piñeiro and Pérez (1998) on the development of the public street lighting and water supply services in the city of Cadiz, among many others. Other particularly relevant studies include *El agua en Cádiz. 75 años de servicios municipalizados* (2002), edited by Juan Ramón Cirici to commemorate the anniversary of the municipalisation of the water supply service in the city of Cadiz, and *Agua, ciudad y territorio*, edited by Juan M. Barragán (1993), which takes a geo-historical approach to studying the supply of water in the city. An analysis of all of these studies has enabled us to understand the difficulties involved in supplying water to the city and the characteristics and development of the supply service throughout the nineteenth century. The information extracted from these sources has been completed with the documentation found in different archives, libraries, and newspaper libraries, such as the press of the period and the *Guías de Cádiz*, from which information about the private companies that participated in supplying water to the city in the period analysed and their characteristics has been drawn. Furthermore, the Municipal Historical Archive of Cadiz conserves the city council minutes records for the period of study and a wealth of dossiers, records and reports that contain miscellaneous information about projects to convey water to the city, the companies responsible for the service, and the general characteristics of the supply, among other details that have also been essential for reconstructing the history of the management and supply of drinking water in Cadiz at the end of the nineteenth century.

The chapter is structured into five sections. After this introduction, the principal difficulties involved in supplying drinking water to the city of Cadiz in the nineteenth century are presented. The third section analyses the different drinking water supply projects developed in Cadiz in the second half of the nineteenth century, focusing not only on their main characteristics, but also on the works necessary to carry them out. The fourth section examines the role played by *The Cadiz Water Works Company Limited* in the supply of water to the city, its principal characteristics, and the specific features of the service. The study ends with some final considerations and conclusions.

2 The Water Supply Problems in the City of Cadiz in the First Half of the Nineteenth Century

The water supply in Cadiz is the result of the evolution of the technical, urban, and economic development of the different civilisations that settled in its territory throughout history. A common feature of the many cultural settlements with their different hydraulic techniques was the limitation of water resources in the area (Pérez and Bestué 2008). In every historical period, the supply of water has been a constant problem in the city of Cadiz, as its geographical location and its dry climate distanced it from natural springs and deprived it from abundant rainfall. In the urban centre, there was only one well of potable water. The rest of the wells in the city contained brackish waters that were unfit for consumption. There were many wells, but the problem resided in the quality of their water. Therefore, it was necessary to find other means of supply. Only rainwater, collected in the tanks installed on the roofs of the houses provided this resource to the citizens of Cadiz. However, these tanks were not a safe solution as a year of scarce rainfall, which was common in the city, led to scarcity in supply of this essential resource (Molina 1993, 151; 2009, 73; Piñeiro and Pérez 1998, 333; Ramos 1992, 36). Therefore, the water was mainly drawn from the springs in the Sidueña or La Piedad valleys (close to the town of El Puerto de Santa Maria) and was tanked and transported in boats

FIGURE 5.1 Remains of the Roman aqueduct at Cadiz
SOURCE: AUTHOR

and then sold by "*aguadores*"[1] to the citizens in barrels; a business subject to all kinds of abuses (Piñeiro and Pérez 1998, 334).

At the beginning of the nineteenth century, several attempts were made to reconstruct the Roman aqueduct (Figure 5.1) that transported water from the Tempul[2] and several studies were conducted on the quantity and quality of its water. However, this was interrupted by the French siege of the city during

1 "*Aguadores*" was the name by which water sellers were known within the city.

2 Tempul is the diminutive term for Valle del Tempe (Barragán 1994, 15), then located in Jerez de la Frontera, although today its area is divided between the municipalities of Algar, Arcos de la Frontera, San José del Valle, and Jerez de la Frontera.

the War of Independence, and it had to resort to water from the wells in the neighbouring town of San Fernando, although it seems that the amounts were insufficient to fully supply the city. After the siege, the transportation of water from El Puerto de Santa Maria resumed (Pérez et al. 2013, 3; Ramos 1992, 36).

On the other hand, it should be noted that during times of drought and greater difficulties, extraordinary measures were adopted, such as the signing of contracts with private owners of springs, collaboration with other towns for the supply or sale of water within the city, which, under normal conditions, could only be carried out by the "*aguadores*". In this respect, the arrival of the railway was particularly significant, as this means of transport reduced the dependency on maritime transport which, during storm periods, enormously hindered the supply to the city (Sánchez 1993, 197–8).

Considering all of this, we can say that the water scarcity and supply problems in Cadiz prevented the adequate and complete urban development of the city. In around the mid-nineteenth century, the city continued to suffer from a shortage of sources and springs and was mainly supplied with the water collected in the tanks constructed on the roofs of all the houses, with the water from the wells located outside the city walls, from which fairly clean water could be drawn, and with the water from El Puerto de Santa Maria, which was the best in the whole region (Rosetty 1871, 155–6). This situation led to the emergence of many improvement initiatives and projects in the years that followed.

3 Drinking Water Supply Projects in the Second Half of the Nineteenth Century

In the 1800s, practically no attempts were made to improve the water supply until the middle of the century. In the 1840s, new projects were developed by José Larios (1842) and Narciso Baltá, the representative of *José Abad y Compañía* (1847). Most of these projects were based on the reconstruction of the old Roman aqueduct, but these types of projects were particularly expensive and the difficulties involved led to their abandonment (Pérez et al. 2013, 3–4).

Subsequently, a series of projects were implemented based on more stable supply structures and a wide range of studies were conducted in the space of a few years. First, in 1855, the British engineer, Thomas White, commissioned by several English companies, conducted a study of a water supply project whereby water was drawn from the Guadalete river beyond La Cartuja and out of reach of the tides. Second, through the Royal Order of 18 April 1861, *Francisco Menoyo y Compañía* was authorised to present water supply studies within the space of one year on the springs of La Piedad in the municipality of El Puerto de Santa Maria. Third, on 7 August of the same year, Guillermo Partington from Madrid offered to bring abundant volumes of quality water within a period of two years. Therefore, taking into account the amount of studies presented in such a short space of time, on 18 September 1861, the Gaceta de Madrid published an edict, establishing a term of four months for the presentation of projects to undertake works to bring water from the springs of La Piedad, the Guadalete, Tempul or Majaceite (in El Puerto de Santa Maria and Jerez de la Frontera) and to construct tanks to store the water in the neighbourhood of San José. Three projects were presented: one by José Mª Fabrés from Madrid, another by *Easton Amot & Hone*, of London (with the Thomas White project) and one by *Francisco Menoyo y Compañía*. The three projects fulfilled the fundamental requisites: to conclude the works within a space of two years, to use the waters from the afore-mentioned springs so as to bring 24,500 m^3 per day to the city and to establish machinery, filters and tanks in the neighbourhood of San José with a storage capacity equivalent to the consumption of four months (Pérez and Román 2015, 161; Piñeiro and Pérez 1998, 334–5).

The high cost of supply, both for private entities and the municipality, led the city council of Cadiz to sign contracts with private entrepreneurs to regulate the supply in the city. On 6 February 1862, Menoya's project was unanimously approved and, although the solution was transitory, it was not sufficient to meet the needs of the citizens of Cadiz (Ramos 1992, 37). Furthermore, Guillermo Partington sought a review and the Ministry of the Interior ruled in his favour with the Royal Order of 13 December 1862, which, in its provisions, ordered the city council to issue a new call for the presentation of projects, delaying the beginning of the works. In compliance with this Royal Order, the engineer Luis de la Escosura was appointed to study the conditions that the selected project had to fulfil. His report indicated that the amount of available water should be 50 litres per inhabitant, that the total budget of the works could not exceed 15,000,000 reales and that the ideal place for the extraction of water was from the springs of La Piedad, in the Sidueña valley, close to El Puerto de Santa Maria and from where this town drew its water with a possible supplementary resource taken from the river Guadalete. The extraordinary meeting approved this project on 26 June 1863, marking the beginning of the problems with the municipality of El Puerto de Santa Maria, as the choice of this place for extracting water would give rise to several lawsuits between the two town councils over the years (Fierro 2002, 51–2; Pérez and Román 2015, 161; Piñeiro and Pérez 1998, 335).

In Cadiz, after many unfinished projects, an agreement was finally reached between the city council and a private individual, Matías del Cacho. Thanks to the records conserved in the Municipal Historical Archive of Cadiz, we have been able to determine that this entrepreneur approached the city council on 5 January 1867 requesting the concession for 99 years for supplying water to the city and its outskirts, including its bay.[3] The contract between the parties was not signed until 4 February 1868 and was endorsed by the Royal Order of 30 July of the same year, giving rise to the constitution of the *Empresa de la traída de aguas de los manantiales de La Piedad a Cádiz*. Matías del Cacho pledged to construct a canal to transfer, if necessary, up to 278 litres per day per inhabitant from La Piedad springs to Cadiz. In the case where the waters of La Piedad were insufficient to meet the needs of the population, the necessary works would have to be carried out to bring resources from other sources. Furthermore, in the area of the city outside of the walls, in the neighbourhood of San José, the entrepreneur was obliged to construct the tanks necessary to store 40,000 m^3 of water and locate them at a height so that they may supply all of the buildings in the highest parts of the city. The concessionaire was also to construct a pipe network inside the city which would run through its streets and squares, facilitating the supply of water to all those buildings requesting it. However, the installation of pipes inside the homes and the costs of conveyance from the general piping system to them were borne by the owners or tenants. A period of three years was set for completing the works which had to begin within two months after the concessionaire was notified of the approval of the contract. At the end of the concession, after the stipulated 99 years, the company was to transfer all the property rights to the city council, with no compensation.

The concession contract also indicated the tariffs that should prevail for the sale of water, establishing the prices for the first 50 years of the concession (Table 5.1) and others for the last 49 years. Furthermore, the concessionaire would provide, free of charge, 162.24 m^3 per day to the city council for public services, with an annual value of 126,000 reales de vellón (hereafter, rv). In the case where the local government needed a higher amount, it would obtain it at a price that was 50 per cent lower than those applicable to private individuals. Finally, it also established that the concessionaire was obliged to make a deposit of 1,000,000 rv, in state debt securities at a consolidated 3%, in order to guarantee the fulfilment of the contract. This amount would be reimbursed at the end of the works, provided that they were carried out under the established terms and conditions; otherwise, the deposit would be lost and, moreover, the

3 Archivo Histórico Municipal de Cádiz – Municipal Historical Archive of Cadiz – (hereafter, AHMC). C.4646. Water supply: contract signed with Matías del Cacho.

TABLE 5.1 Prices for the sale of water to individuals during the first 50 years of the concession[4]

Total daily sales in the city	Price of the daily water real (in reales de vellón)
1 to 500 water reales (1,622.4 m^3)	15
501 to 1,000 water reales (3,244.8 m^3)	14
1,001 to 1,500 water reales (4,867.2 m^3)	12
1,501 to 2.000 water reales (6,489.6 m^3)[a]	10

a From 2,001 water reales and above, one rv was discounted for every 500 water reales which increased the total sales.

SOURCE: OWN ELABORATION BASED ON AHMC. C.4646. WATER SUPPLY: CONTRACT SIGNED WITH MATÍAS DEL CACHO

contract would be cancelled and the city council would be the owner of the works that had been carried out until that point.[4]

A few months later, on 12 June 1868, after the deposit was made with the city council, the works on La Piedad springs began (Rosetty 1869, 49). The difficulties exceeded the concession company's investment and technical expertise, obliging it to repeatedly request extensions in order to finish the works.

In the summer of 1870, Matías del Cacho presented a report on supplying water to the city, analysing the expenditure and works that had been undertaken until then, and also those that had yet to be carried out.5 The report indicated that the expenditures made until 31 July 1870 for works, materials, studies, inauguration, etc., amounted to 2,500,000 rv. On the other hand, the works still to be undertaken and the material necessary for executing the project amounted to 11,059,462 rv (Table 5.2). Therefore, a company was created with a capital of 16,000,000 rv, as without this amount, it would have been impossible to meet the needs of the water supply project. The objective of Matías del Cacho was to obtain the capital necessary to undertake the works and be able to fulfil the concession contract for supplying water to the city.

On the other hand, it should be noted that from 25 July 1870, another large water supply company began operating under the name of *La Cruz*, established

4 For the last 49 years of the concession, the price for individuals increased to 7 rv per day and water real, if the total sales were between 1 and 4,000 water reales, and 5 rv if they exceeded 4,000 water reales.

5 AHMC. C.4646. Report on the water supply to Cadiz and the Company Project. 1870.

TABLE 5.2 Capital necessary for executing the water supply project, 1870

Concept	Amount (in reales de vellón)
Works carried out and other expenditure from 1 January 1868 to 31 July 1870[a]	2,500,000
Studies by Palacios y Pau	80,000
Studies by the engineer Mr. Mouchez	250,000
Inauguration	80,000
Fittings	20,000
Instruments	4,000
Property, jobs done, and materials gathered	1,110,760
Material	72,000
General costs	543,240
Deposit	340,000
Tasks and materials necessary for conclusion	11,059,462[b]
Main pipe network from the Cadiz deposit	5,329,836
Pressure pipes to raise the water	259,120
Piping inside the city	248,000
Piping to concentrate the water	158,720
Pipe system for the connection in El Puerto de Santa Maria	198,400
Repair of the old pipes of La Piedad	60,000
Ten cranes for Cadiz and their placement	48,000
Steam machine with its corresponding pumps	400,000
Customs duties	600,000
Building for housing two machines	280,000
Termination of the tanks for 15,000 m3 of water	400,000
A central well	80,000
Works for water collection	1,200,000
Purported benefits for contractors	1,364,116
Contingencies	433,054
Organisation and administration costs during construction	300,000

a These do not include the studies and expenditures undertaken prior to obtaining the concession, which amounted to approximately 1,200,000 rv, and were borne by the concessionaire.

b The total of the pending works amounted to 11,059,246 rv. However, there was an error in the budget and the total sum was taken as 11,059,462 rv, representing an excess of 216 rv.

TABLE 5.2 Capital necessary for executing the water supply project, 1870 (*cont.*)

Concept	Amount (in reales de vellón)
For the interest of one year, on average, the maximum calculated for concluding the works and beginning operations	1,560,000
Reserve fund and contingencies	580,538
Total	**16,000,000**

SOURCE: OWN ELABORATION BASED ON AHMC. C.4646. REPORT ON THE WATER SUPPLY TO CADIZ AND THE COMPANY PROJECT. 1870

by José F. De Prado, offering considerable advantages to the population. De Prado had the Fuenterrabía spring in El Puerto de Santa Maria at his disposal, which had excellent waters. In order to resolve the drought problems afflicting the city, De Prado decided to implement several improvements. To do this, in December 1869 he approached the city council requesting the assignment of two of the warehouses under the quay wall in which to install a tank that would contain enough water so as ensure that there would never be a scarcity. Considering the immense advantages to be gained from this project, the municipal council agreed to allow De Prado to use these warehouses free of charge for a certain number of years. This entrepreneur committed to supplying the city council in times of drought with fifteen free water tanks to supply the needy classes and, if more was needed, to provide a further fifteen at the low price of six reales each.

After approving the contract and the cession of the warehouses, several obstacles delayed the delivery, which did not take place until the beginning of June. The works to adapt the warehouses began on this date and were finished in just forty days. This initiative was well received in the city due to the scarcity of rainfall which the population had experienced in the preceding years and due to the low prices of the water supplied.

The company transported the water from the La Piedad and Fuenterrabía springs with its own vessels. This represented an advantage over any other competitor. A tank of water from El Puerto emptied into the roof-top tanks had a price of 17 reales and one from the Fuenterrabía spring cost 15 reales. In addition to these fairly economical tariffs for the general public, the public coffers greatly benefited as the city council was able to supply the poor classes with water for free with the number of "*pipas*" that it received each day (Rosetty 1871, 155–6; 1873, Advertising section, 53).

Returning to the company of Matías del Cacho, it should be indicated that, despite the attempts to finalise the project of transferring water to the city and given that the difficulties continued, in September 1871, the concessionaire saw the need to modify some of the conditions of the contract, as time and experience had revealed the limitations of the stipulations initially established. The city council commissioned the engineer Luis de la Escosura to conduct a study, which, after confirming the reality of the situation, brought about the modification of the clauses of the contract referring to the amount of water to supply, the execution and duration of the works, the installation of the tanks and the tariffs for the sale of water. After the restructuring of the contract, first, the amount of water that the concessionaire was obliged to supply to the city was reduced to 100 litres per day per person. Second, the construction of the tanks in the neighbourhood of San José was cancelled and all the works to be carried out by the concessionaire had to be finished within a period of eighteen months, counting from the day on which these modifications were approved; 2 October 1871. The tariffs for the sale of water were also modified, adding prices for the sale in perpetuity (Table 5.3) to the prices for monthly or annual subscription (Table 5.1).[6]

On 30 October 1871, despite the restructuring of the contract and the attempts to improve the supply of water to the city, the impossibility of fulfilling all the stipulations led Matías del Cacho to agree to the transfer of the concession and all of the rights, works, land, machinery, materials and other

TABLE 5.3 Prices for the sale of water in perpetuity

Total daily sales in the city	Price of the daily water real (in reales de vellón)
From year 1 to year 5	24,000
From year 6 to year 10	20,000
From year 11 to year 20	18,000
From year 21 to year 40	16,000
From year 41 to year 60	14,000
From year 61 to year 80	12,000
From year 81 to year 99	10,000

SOURCE: OWN ELABORATION BASED ON AHMC. C.4646. WATER SUPPLY: CONTRACT SIGNED WITH MATÍAS DEL CACHO. 22 SEPTEMBER 1871

6 AHMC. C.4646. Water supply: contract signed with Matías del Cacho. 22 September 1871.

properties to the London-based company *The Cadiz Water Works Company Limited*.[7]

4 *The Cadiz Water Works Company Limited* and Its Role in the Service of Supplying Water to the City of Cadiz

The Cadiz Water Works Company Limited (Figure 5.2) began to supply water to the city of Cadiz in the 1870s. The deed for the transfer of rights from Matías del Cacho and the new concession to the English company was signed on 17 April 1872 (Fierro 2002, 52–3). However, the transfer was not immediate and the supply of the water to the city continued to suffer from deficiencies as the works necessary to improve it were not being undertaken. The situation was such that the *Liga de Contribuyentes de Cádiz* (Taxpayers' Association) formally requested the city council to cancel the water supply contract of Matías del

FIGURE 5.2 Emblem used by the company *The Cadiz Water Works Company (Limited)*
SOURCE: AHMC. C.4646. CADIZ CITY COUNCIL RECORDS. WATER COMMISSION. E.47-BOX 27. 1873

7 AHMC. C.4646. City Council records. 4 September 1872.

Cacho on the grounds that the concessionaire was not duly fulfilling the commitments acquired, which was seriously harming the city and its inhabitants.[8]

Finally, *The Cadiz Water Works Company Limited* took over the works necessary to supply the city with water in April 1872. The company's registered office was in London, at 4 Skinners Place, Size Lane and the offices in Cadiz were located in Calle de la Cruz de la Madera. Initially, the board of directors was made up of the chairman, William Hawes, the secretary, Benjamin G. Kinnear, and the bankers, *Robarts Lubbock & Co*, with the company being directed by Horacio Alcon Beranger. However, in the following years, certain changes were made. In 1873, John Arthur Wright was appointed as the managing director of the company and, in 1874, C. S. Collard was brought in as secretary. Consequently, Kinnear became a member of the board, together with Alcon and Wright. Furthermore, on 8 July 1873, the company named these latter two members as representatives of the company to act on behalf of the company both in Spain and abroad.[9]

At that time, nine offices had been established in the city for the sale of water by the company, located in the streets of Albenda, Bendición de Dios, Botica, Consolación, Desamparados, Diego Arias, Doblones, San José and in the square of the Cathedral. However, in the following year, in 1875, the offices for direct sales were reduced to just three; those in the streets Albenda, Doblones and San José. Moreover, four taps were installed on the extension of the quay to serve the bay and offices were opened in El Puerto de Santa Maria, specifically in Calle Larga 74, where Valentín G. Bell was chief engineer and John Arthur Wright was the functional engineering director, both representatives of the company (Rosetty 1875, 256 and 374; 1876, 264).

At this point, it is necessary to examine the works undertaken by the British company in Cadiz, which were not exempt from the difficulties arising from the need to acquire land which was essential for the full development of the project. The works were carried out by the contractor *Barnett & Gale* of London and were executed under the direction of Guillermo Forrect. To do this, the water company acquired the Tiro orchard where many springs were located. Furthermore, it traded 12,000 m^2 of the plots of Verdugo with the competent authorities for others of the afore-mentioned orchard and acquired the ownership of the lands of the Sierra de San Cristóbal (commonly known as Buenavista), where two tanks were built, each with a capacity of 15.000 m^3,

8 AHMC. C.4646. Copy of the proposal presented to the City Council of Cadiz by the *Asociación* or *Liga de Contribuyentes de Cádiz*, regarding the cancellation of the water conveyance contract of Matías del Cacho. 11 August 1872.

9 AHMC. C.4646. City Council records. 10 September 1873; Rosetty 1873, 255; 1874, 261; 1875, 256.

together with buildings in which to house the workshops. It also acquired other land in the Sidonia valley to drill a well with a depth of 18 metres and a diameter of five metres. Furthermore, it leased an area of land in Manchones de Hierro and a series of wells in order to continue the works unhindered. In addition to these contracts, an area of land was expropriated in the lowlands of El Puerto de Santa Maria to construct a railway line to serve the company, which connected with the general Seville-Jerez-Cadiz line. Finally, large steam machines were purchased to extract the water and 2,000 tonnes of pipes to transport it, which were laid alongside the railway line. In this respect, Valentin G. Bell, who had been named representative of the water supply company on 26 March 1872,[10] signed a contract with the *Compañía de los ferrocarriles de Sevilla-Jerez-Cádiz* on 13 April 1873 for the laying of pipes or an iron aqueduct along the whole length of the railway line in order to transport the water from La Piedad or Sidonia valley to Cadiz, provided that it did not interfere with the railway service. To do this, the water company would pay the railway company 200,000 rv. The result of all these investments would be 800,000 m^3 of water per day, that is, 16,000 *botas,* an amount that infinitely exceeded that calculated initially for the land and wells described (Rosetty 1873, 185–6).[11]

On the other hand, during the summer of 1873, certain problems arose with the town council of El Puerto de Santa Maria, which interrupted the plan proposed for June to September of that year (Rosetty 1875, 194). The local government had already attempted to suspend the works on several occasions and even requested the annulment or expiry of the contract of Matías del Cacho some years earlier, as no works had been carried out. However, as we have seen, the arrival of the English company enabled the works to resume, but as they progressed, the opposition from the local government of El Puerto increased. It alleged all kinds of problems, such as defects in the form of the contracts and protests regarding the ownership of the land where the works had to be undertaken. In this respect, the works were paralysed in June until the water company accredited the authenticity of its plans and established that all the works were being carried out in accordance with them. After resolving this delay, the local government of El Puerto suspended the works on the railway line, even though the approved plans and the order of the Ministry of Development had been submitted. The local government of El Puerto not only sought to ensure that all the projects were being executed according to the plans, but also that they were not, either then or in the future, harmful to the old works from where the waters for the city were drawn. It also sought to verify the amount of available

10 AHMC. C.4646. City Council records. 26 March 1872.

11 AHMC. C.4646. City Council records. 13 April 1873.

water in La Piedad valley. Furthermore, it believed that the municipality of Cadiz was stealing or adulterating the water from La Piedad to supply itself. These demands and this systematic opposition only served to seriously damage the English company. Therefore, after taking the appropriate legal action, the company approached the city council of Cadiz to request its intercession and for it to mediate with the local government of El Puerto and with the provincial authorities in order to ensure the continuation of the water supply works.[12]

Furthermore, in August it requested the city council of Cadiz to exempt from taxation the material that was imported from abroad used to carry out the water supply works. This was approved, given the difficulties suffered by the company and the desire of the councillors to see the city supplied with drinking water.[13]

After the drawbacks had been overcome and the problems with the neighbouring municipality of El Puerto de Santa Maria had been resolved, the works continued at a good pace until their conclusion in May 1874. Over the following months, the water reached Cadiz, Puerto Real and El Puerto de Santa Maria, which were the towns that had to be supplied (Matés-Barco 2020, 19). With respect to Cadiz, the *Guía de Cádiz* of 1875 refers to the arrival of the water to the city in the early hours of 10 July:

> [...] we can confirm the material fact that the crystalline flows of the pure, abundant and exquisite water from Sidonia's valley have been supplied to this city since the early hours of the morning of the 10 July, when the first official test was carried out on the embankment of the railway station in the presence of his Excellency The Governor of the province, the Water Commission and some other members of the City Council, the representatives of the supply company which has made an immense and invaluable contribution to our Cadiz, Mr. Juan Arturo Wright and his Excellency Sr. D. Horacio Alcon, and a large crowd, which, from the town hall square followed the afore mentioned to the embankment and attended the event. The public broke into spontaneous applause and cheered the company when the engineer Mr. Wright opened the key allowing the water to flow. The dream of many years of all of the authorities and of those who have devoted great efforts to this city and who now see this dream come true. This is thanks to those who, despite the eventful and unfavourable situation in which the country has been immersed, subject to

12 AHMC. C.4646. City Council records. 20 August 1873.

13 AHMC. C.4646. City Council records. 25 August 1873.

> horrific anarchy and as a result, torn apart by the regrettable civil war which devours us, did not hesitate to undertake an arduous and costly venture, which has consumed and continues to consume large amounts of capital. (Rosetty 1875, 193–4)

With respect to the conditions of the supply contracts to private individuals, on 15 September 1873, the company published a prospectus which outlined the general terms of the contracts signed with the city council and the systems that were adopted for the supply to homes (Rosetty 1874, 65–9). Moreover, they indicated the advantages and convenience of receiving water directly in the home. This prospectus informed that the company had adopted two systems: the supply of water using a meter and without a meter, with a pipe of a specific diameter. They were both based on the precept that consumption had to amount to a minimum of half a cubic metre per day, in accordance with the agreement with the Cadiz City Council. The prices varied depending on the level of overall consumption (Table 5.4); they fluctuated from 4.60 rv per cubic metre when sales did not exceed 1,622 m^3 to 2.76 when they exceeded 6,491 m^3. Overall consumption would be published each semester in the Official Bulletin of the Province to inform the consumers.

On the other hand, the monthly leasing prices and those of the repairs of the meters ranged from 6 to 70 rv, depending on the size and supply of the meter (Table 5.5). All the payments would be made in advanced monthly instalments and, in the case of the supply with a meter, the excess indicated by the meter would be settled the following month. Furthermore, the water outlet would always be verified by the company at the consumer's expense, although the distribution could be carried out by the consumer but using the operators and plumbers authorised by the company.

TABLE 5.4 Tariffs of *The Cadiz Water Works Company, Ltd.*

Public subscription	Daily price per m^3 (reales de vellón)
1 to 500 water reales (1,622 m^3)	4.60
501 to 1,000 water reales (3,245 m^3)	4.32
1,001 to 1,500 water reales (4,868 m^3)	3.60
1,501 to 2,000 water reales (6,488 m^3)	3.00
2,001 water reales and above (6,491 m^3)	2.76

SOURCE: ROSETTY 1877, 59

TABLE 5.5 Tariffs of *The Cadiz Water Works Company, Ltd.* for monthly leasing and repairs of the meters

Size of the meter	Daily supply (m^3)	Monthly lease, including repairs (reales de vellón)
10 mm	6	6
12 ½ mm	12	9
20 mm	24	12
25 mm	36	15
35 mm	72	20
50 mm	144	25
75 mm	324	40
100 mm	600	70

SOURCE: ROSETTY 1877, 59

Moreover, the consumer could not give, waste, or supply water to anyone outside of their household, or apply the water to uses other than domestic uses, being responsible for the misdemeanours of their dependents and servants. The contracts were established on a yearly basis and could be extended every six months. Finally, the company had the right to suspend the water supply with no warning in the case of a breach of the conditions of the contract or a failure to pay and the service would not be continued until the payment of the debt and of the cost of opening and closing the valves or the reestablishment of the corresponding pipes.

However, despite the efforts and investments made in the first half of the 1860s, the results were not entirely satisfactory, as the company was only able to guarantee a supply of 2,162 m^3 per day for the capital (Pérez et al. 2013, 5). In addition to the insufficient water, there were also problems related to its quality as the water that was received was brackish. The supply was also irregular due to constant fractures in the network. After the report submitted by the engineer Luis de la Escosura, the city council detected serious differences with the original proposal. There was a large volume of water, but improvements were required to exploit it fully. Therefore, the company was given an extension of eight months to undertake the necessary works and excavations, which, once again, required an enormous effort and large investments (Fierro 2002, 54–5).

Thus, in general terms, the company gave a vigorous boost to the drinking water conveyance works in the city of Cadiz, investing large amounts of capital to be able to meet the needs of the citizens (Rosetty 1879, 212–3).

In order to facilitate the transport of the large pieces of machinery and pipes, the company constructed a railway siding. After obtaining the easement from the *Compañía de los ferrocarriles de Sevilla-Jerez-Cádiz* for all the land, bridges and constructions of this company, the aqueduct, which was composed of an iron pipe with a large diameter, ran from the tanks to the capital across a distance of 38.807 km. It followed the line through El Puerto de Santa Maria and Puerto Real, it crossed the area between San Fernando and El Arsenal de la Carraca and passed through the neighbourhood of San José outside of the walls and the land of the railway station to the principal quay, crossing the walls of the city at the Puerta del Mar.

The company also supplied the city council with several fountains distributed across the whole of the city. They were made of iron and had four taps, located in the squares of the Cathedral, Cuatro Torres, Méndez Núñez, Merced and Rey Don Alfonso XII and in the streets of Plocia and Porlier. There were also four in San Bernardo Street and another four in the Paseo del Sur. In the neighbourhood of San José outside of the city walls, there were another two fountains, one in the square with the same name and another in the area called Buena Vista in La Primera Aguada. A decorative fountain was also installed in the centre of the Mina square, one at each end of the Paseo de las Delicias, one in the centre of the gardens of the Merced square and another in the Reina square. Those of the gardens of the Paseo de las Delicias, the one in Plaza Fragela, the two in the Alameda de Apodaca and the one in Tinte street were also supplied with drinking water. There were also other fountains in the Chapter Houses, in the Municipal Children's Asylum, in the provincial civilian hospital and the Civil Guard Barracks House. The 162 m^3 of water provided to the city each day was distributed between these public services, in accordance with the conditions of the contract signed with the supply company. It had also installed 61 irrigation hydrants in different parts of the municipality and another four taps in the quay in order to serve the needs of the vessels in the bay (Figure 5.3). Finally, column dispenser fountains were installed in the squares of the Cathedral, Rey Alfonso XII and San Francisco and in Plocia, San Bernardo and Tomás Isturiz streets for public use (Rosetty 1879, 213).

However, the city's water supply problem persisted; the works had not generated the expected effects, as the necessary supply was not achieved, and an extension of the springs was required. The company could not assume the continuous difficulties and the high and constant investment that the business required. Therefore, it declared the suspension of payments, bankruptcy and, finally, its auction and liquidation in 1880 (Fierro 2002, 54–5; Rosetty 1881, 304).

This company constitutes a clear example of the high level of instability that characterised this sector in Spain. There were many companies that survived

FIGURE 5.3 Advertisement of *The Cadiz Water Works Company* (*Limited*) in the *Guías de Cádiz* in the last third of the nineteenth century
SOURCE: ROSETTY 1880, ADVERTISING SECTION, 82

in the water market for just a few years, as they were unable to assume the difficulties of a sector that was developing rapidly (Pérez-de-la-Cruz et al. 2020). Moreover, similarly to Cadiz, in other cities, such as Barcelona, Seville, Jerez de la Frontera, A Coruña, and Alicante, large foreign investors or entrepreneurs were more willing to invest in the sector, obtaining many of the concessions for the supply of water. Foreign capital or investment (mainly from Britain, France, and Belgium) also contributed to the development of the sector, thanks to the arrival of new knowledge, technologies, and industrial material. However, this phenomenon was not exclusive to the water supply companies, as it was the result of a trend that took place in certain public services, such as the railways and, most of all, the first gas, tram, and electricity companies. Therefore, in this respect, the city of Cadiz constitutes a good example of the trends that prevailed in the early phases of the sector (Matés-Barco 2014, 52–4; 2021, 145–7).

The solution to the water conveyance and supply problems came through a rich landowner from Cadiz, Diego Fernando Montañés y Álvarez, who, after his death in 1874, left in his will the legacy necessary to undertake the two public works of greatest interest for the city: the supply of drinking water and the reform of the port.[14]

The executors of the will of Montañés began their task by consulting the capital invested by the English company in the infrastructure works. The company reported that until 2 February 1877 it had spent £558,500. Part of the population considered that, despite the investments carried out, the deficient water supply was the exclusive fault of the British company and that it should provide the pertinent solutions to the supply problem. For others, the inefficiency repeatedly exhibited by this company highlighted the need to make use of the legacy of Montañés. Within this context, the city council did not adopt any measures to mitigate the supply problems of the city, while the executor of the will waited for indications in order to grant the aid. In 1880, *The Cadiz Water Works Company Limited* decided to sell its properties and the executor of Montañés will participated in the auction in order to obtain the company, under the auspices of the local government. However, after several incidences, the company was sold to the English entrepreneur John Syer, who purchased it for his French clients (Matés-Barco 1997, 115; Ramos 1992, 38).

The operation involved the transfer of all of the land, water supply works and other works of the company under construction in the city of Cadiz and its surrounding areas, together with all of the galleries, tanks, pipes, machines, instruments and accessories related to them, except for the loans in its favour. The price of the transfer was £35,200 in total (875,000 ptas), £30,200 for the

14 AHMC. C.5636. Third copy of the closed testamentary clauses of Diego Fernando Montañés y Álvarez and of the proceedings carried out for the reading and registration of the will.

immovable property and £5,000 for the movable assets.[15] The immovable assets included in this sale together constituted an estate called "The conveyance of waters to Cadiz", in the town of El Puerto de Santa Maria, made up of the following properties:

1. A 12,000 m^2 area of land, which formed part of the Verdugo or Esquivel gorge in the valley of Sidonia.
2. An area of 360 m^2 of land, also in the Sidonia valley, specifically in a place known as Caño de la Vega.
3. A piece of land of 8 hectares, 81 ares and 29 centiares in the San Cristóbal Sierra.
4. An area of fallow land of 3 ares in the San Cristóbal Sierra, in the aforementioned valley.
5. An area of land of 92 ares that formed a belt from the Verdugo gorge, joining with the Seville-Jerez-Cadiz railway line, forming part of the Dehesa de la Vega.
6. Another belt of land of 34 ares, running from east to west of the estate of Antonio Quirós.
7. An area of land of 2 hectares, 20 ares and 40 centiares in Manchón and Huerta de Hierro.
8. An orchard called Tiro in the Sidonia valley with 6 hectares, 5 ares, 74 centiares and 60 miliares. This estate had two springs with an abundant supply of drinking water, a farmhouse, and a grove.
9. A plot of land of one hectare, 67 ares and 70 centiares in the Tiro valley.
10. An area of land of 4 hectares, 90 ares and 76 centiares with a house, well and water wheel, in the rest of the orchard and Manchón de Hierro in the Sidonia valley.
11. The easement of aqueducts from the concentration wells located on the land of the former water company through the higher Verdugo gorge, common lands of the neighbours and the San Cristóbal Sierra, to the tanks located in the same sierra and from there to tanks in the other afore-mentioned sierras, lands and gorges between the Madrugador olive grove, El Pinar de Manrique and the Dehesa de la Vega, joining the Seville-Jerez-Cadiz railway line between kilometres 121 and 122. This easement was constituted in favour of the former water company by the Cadiz city council in accordance with the Water Law, with the conditions that the

---

15 AHMC. C.5636. Cadiz City Council. The liquidators of the company *The Cadiz Water Works Company Limited* and representatives of its bonds and obligations to the new company *The Cadiz Waterworks Limited*, before Esteban Paullada y Moreno, Notary of the Illustrious Bar Association of Seville. 21 March 1883.

pipe would be covered and ditches of no more than 100 metres would be dug so as not to hinder the passage of cattle or interrupt the supply of water to this city, according to the deed signed on 29 September 1873.

12. Another easement of aqueducts through all the land, viaducts, bridges, and stations of the *Compañía de los ferrocarriles de Sevilla-Jerez-Cádiz* from kilometre 121 to the station of Cadiz and the quays of the Puerta del Mar of this city.
13. The sale also included an olive grove called Caletilla, located in the Sidonia valley with an area of nine squared hectares, 81 ares, 58 centiares and 46 decimetres.

On these properties, houses had been constructed for rooms and for machinery, together with workshops, warehouses, wells, tanks, galleries, and aqueducts; all of this was recorded in the Property Register as a single estate.

With all the assets, in January 1881, the new owners constituted a new company in London with the name *The Cadiz Waterworks Limited*, which sought to undertake the supply of water to the city in the following years (Rosetty 1882, 282; Advertising section, 195; 1883, 283).

5 Final Considerations

The city of Cadiz has traditionally been characterised by its limited water resources; its geographical location and its dry climate distance it from natural springs and deprive it from the collection of abundant rainfall. Therefore, the supply of water has been a constant problem in the city throughout history. In order to find a solution, several projects were proposed during the nineteenth century, when the private initiative boosted the development of this sector.

Among these private initiatives, one that was particularly significant was that of the London-based company *The Cadiz Water Works Company Limited*, whose involvement in the supply and management of water in the city was essential for the improvement of the service between 1872 and 1880. The short life of this company constitutes a clear example of the high level of instability that characterised this sector in Spain. Many companies survived in the water market for just a few years, as they were unable to assume the difficulties of a sector that was developing rapidly.

During the eight years of its operations, *The Cadiz Water Works Company Limited* made an enormous effort and invested large amounts to acquire immovable assets and to undertake many works to improve the water supply service of the city. It also provided the city council with several fountains and irrigation hydrants distributed across the city and installed taps on the quay to meet the needs of the vessels in the bay. In general terms, thanks to these investments, the London-based company boosted the drinking water conveyance works in the city of Cadiz and was able to meet the needs of its citizens.

However, the activity of this company was not an isolated case, as in other cities, such as Jerez de la Frontera, Seville and Barcelona, large foreign investors or entrepreneurs contributed new knowledge, technology, and industrial materials and, therefore, to the development of the sector, obtaining many of the concessions for the supply of drinking water. Therefore, thanks to this initiative, the city of Cadiz constitutes a clear example of the trends that prevailed in the early phases of this sector.

However, despite the efforts of the company analysed in this study, the city's water supply problem persisted; the works had not generated the expected effects, as the necessary supply was not achieved, and an extension of the springs was required. Considering this situation, the company could not overcome the continuous difficulties or assume the high and constant investment that the business required. Therefore, it declared the suspension of payments, bankruptcy and, finally, its auction and liquidation in 1880.

The company and all its assets were sold to a group of French entrepreneurs who, in January 1881, constituted a new company in the city of London under a very similar name to its predecessor, *The Cadiz Waterworks Limited*, which sought to undertake the supply of water to the city in the following years.

6 Manuscript Sources

Archivo Histórico Municipal de Cádiz – Municipal Historical Archive of Cadiz – (AHMC): Section 18.13. Water and lighting. Boxes C.4646, C.5636.

References

Barragán Muñoz, J. M. (ed.). 1993. *Agua, ciudad y territorio. Aproximación geo-histórica al abastecimiento de agua a Cádiz*. Cadiz: Universidad de Cádiz and SMAES.

Barragán Muñoz, J. M. (ed.). 1994. *Aguas de Jerez. Tempul: entre el medio natural y la técnica hidráulica*. Vol. 3. Jerez de la Frontera (Cádiz): Aguas de Jerez, Empresa Municipal, S.A.

Cirici Narváez, J. R. (ed.). 2002. *El agua en Cádiz. 75 años de servicios municipalizados*. Cadiz: Aguas de Cádiz S.A. and Excmo. Ayuntamiento de Cádiz.

Fierro Cubiella, J. A. 2002. "El abastecimiento de agua a Cádiz a través de la historia." In *El agua en Cádiz. 75 años de servicios municipalizados*, edited by J. R. Cirici Narváez, 37–61. Cadiz: Aguas de Cádiz S.A. and Excmo. Ayuntamiento de Cádiz.

Matés-Barco, J. M. 1997. "Las sociedades anónimas de abastecimiento de aguas en Andalucía: una primera aproximación." *Boletín del Instituto de Estudios Giennenses* 167: 103–30.

Matés-Barco, J. M. 2014. "Las empresas concesionarias de servicios de abastecimiento de aguas potables en España (1840–1940)." *TST. Transportes, Servicios y Telecomunicaciones* 26: 36–67.

Matés-Barco, J. M. 2017. "El servicio público de abastecimiento de agua en España (siglos XIX y XX). El proceso de acumulación de competencias de los ayuntamientos." *Revista Brasileira de História & Ciências Sociais* 9 (18): 36–57. https://doi.org/10.14295/rbhcs.v9i18.448.

Matés-Barco, J. M. 2020. "El suministro de agua (siglos XIX y XX). Una historia discontinua." *Andalucía en la historia: La ciudad moderna. Infraestructuras en red* 68: 14–21.

Matés-Barco, J. M. 2021. "Public Services in Spain: The Role of Water Supply Companies." In *Companies and Entrepreneurs in the History of Spain. Centuries Long Evolution in Business since the 15th Century*, edited by M. Vázquez-Fariñas, P. P. Ortúñez Goicolea, and M. Castro-Valdivia, 135–59. Palgrave Studies in Economic History. Cham: Palgrave Macmillan. https://doi.org/10.1007/978-3-030-61318-1_9.

Molina Martínez, J. M. 1993. "El agua en Cádiz durante la Edad Moderna." In *Agua, ciudad y territorio. Aproximación geo-histórica al abastecimiento de agua a Cádiz*, edited by J. M. Barragán Muñoz, 137–80. Cadiz: Universidad de Cádiz and SMAES.

Molina Martínez, J. M. 2009. "Agua y ciudad. El abastecimiento de agua en Cádiz hacia 1812." In *Agua, Territorio y Ciudad. Cádiz de la Constitución. 1812*, edited by J. M. Fernández-Palacios Carmona, 71–81. Sevilla: Consejería de Medio Ambiente, Junta de Andalucía.

Pérez-de-la-Cruz, F. J., A. Trapote-Jaume, J. Melgarejo-Moreno, and J. Chazarra-Zapata. 2020. "A Century of Water Supply Companies and Their Influence on the Development of Spanish Society (1842–1942)." *Water* 12, 2634. https://doi.org/10.3390/w12092634.

Pérez Marrero, J., and I. Bestué Cardiel. 2008. "Avance del estudio hidráulico del acueducto romano de Gades." *IV Congreso de las Obras Públicas en la Ciudad Romana. CITOP.* Lugo. Accessed 30 March 2022. http://www.traianvs.net/pdfs/2008_cadiz01.pdf.

Pérez Serrano, J., A. Román Antequera, F. P. Villatoro Sánchez, and J. M. Arévalo Santiago. 2013. "El abastecimiento de aguas en la Bahía de Cádiz." In *Claves del mundo contemporáneo. Debate e investigación: actas del XI Congreso de la Asociación de la Historia Contemporánea*, edited by T. M. Ortega López, and M. A. Del Arco Blanco, 1–12. Granada: Comares.

Pérez Serrano, J., and A. Román Antequera. 2015. "Conflictos por los recursos en la Bahía de Cádiz: la lucha por el control del agua (1850–1935)." *Revista de la Historia de la Economía y de la Empresa* IX: 155–176.

Piñeiro Blanca, J. M., and J. Pérez Serrano. 1998. "La diversificación de funciones de la ciudad europea industrial: los servicios públicos de alumbrado y abastecimiento de aguas en la ciudad de Cádiz (siglos XIX y XX)." In *Panfletos y materiales. Homenaje a Antonio Cabral Chamorro, historiador (1953–1997)*, edited by M. Aguilar Villagrán, J. Cabral Bustillos, A. Fuego García, J. García Cabrera, and C. Orellana González, 333–62. Sevilla: Imprenta Pinelo.

Ramos Santana, A. 1992. *Cádiz en el siglo XIX. De ciudad soberana a capital de provincia.* Madrid: Sílex Ediciones.

Rosetty, J. 1869. *Guía de Cádiz, El Puerto de Santa María, San Fernando y el Departamento para el año 1869.*

Rosetty, J. 1871. *Guía de Cádiz, El Puerto de Santa María, San Fernando y el Departamento para el año 1871.*

Rosetty, J. 1873. *Guía de Cádiz, El Puerto de Santa María, San Fernando y el Departamento para el año 1873.*

Rosetty, J. 1874. *Guía de Cádiz, El Puerto de Santa María, San Fernando y el Departamento para el año 1874.*

Rosetty, J. 1875. *Guía de Cádiz, El Puerto de Santa María, San Fernando y el Departamento para el año 1875.*

Rosetty, J. 1876. *Guía de Cádiz, El Puerto de Santa María, San Fernando y el Departamento para el año 1876.*

Rosetty, J. 1877. *Guía Oficial de Cádiz, su Provincia y Departamento. 1877.*

Rosetty, J. 1879. *Guía Oficial de Cádiz, su Provincia y Departamento. 1879.*

Rosetty, J. 1880. *Guía Oficial de Cádiz, su Provincia y Departamento. 1880.*

Rosetty, J. 1881. *Guía Oficial de Cádiz, su Provincia y Departamento. 1881.*

Rosetty, J. 1882. *Guía Oficial de Cádiz, su Provincia y Departamento. 1882.*

Rosetty, J. 1883. *Guía Oficial de Cádiz, su Provincia y Departamento. 1883.*

Sánchez Álvarez, A. M. 1993. "Los siglos contemporáneos: una etapa clave del abastecimiento moderno gaditano." In *Agua, ciudad y territorio. Aproximación geo-histórica al abastecimiento de agua a Cádiz*, edited by J. M. Barragán Muñoz, 181–241. Cadiz: Universidad de Cádiz and SMAES.

CHAPTER 6

The Improvement of the Water Supply Infrastructures in Antequera, 1818–1932

José Escalante-Jiménez, Mercedes Fernández-Paradas and Carlos Larrinaga

Abstract

This chapter addresses the principal initiatives taken by the local government of Antequera in order to improve the water supply between 1818 and 1932, within the context of the modernisation process of the municipal public services. These projects were characterised by the lack of continuity over time. The first, dating back to 1818, failed. The second, implemented between 1865 and 1867 gave rise to a substantial improvement in the water supply. The third and largest project, developed between 1929 and 1932 sought to fully introduce the modern water supply system.

Keywords

Andalusia – Antequera – public services – urban services – water supply

1 Introduction[1]

This chapter constitutes a first analysis of the attempts of the local government of Antequera to promote infrastructures aimed at improving the water supply between the beginning of the 1800s and the 1930s. After this introduction, the first section describes the principal characteristics that made the medium-sized municipality of Antequera unique within the context of Andalusia, a region of

1 This chapter forms part of the results of the ERDF-University of Jaen-1381621 project (2021–2022) entitled "Sustainable management of public services: Water in Andalusia (1800–2020)", financed by the Department of Economic Transformation, Industry, Knowledge and Universities of the Regional Government of Andalusia and ERDF funds, and the R+D+I project "Gas in Latin Europe: a comparative and global perspective (1818–1945)", PID2020-112844GB-I00, financed by the Ministry of Science and Innovation of the Spanish Government and ERDF funds.

 | DOI:10.1163/9789004679634_007

Southern Spain. The next section provides an overview of the changes in the urban layout and public services. This is followed by a section that describes the most noteworthy initiatives for improving the drinking water supply infrastructures. The chapter closes with the main conclusions.

2 Antequera, a Unique, Medium-Sized Town

The capitalist society in Antequera took shape between 1835 and 1880. At its peak, there was a dominant group made up of large landowners of the bourgeoisie and nobility. In the next phase, this group was replaced by a bourgeoisie engaged in industrial, agricultural and commercial activities, which opted to diversify its investments. The members of this group were connected to the dominant leaders through family and economic ties. Below them, there was a small-sized middle class and finally, the largest part of the population, more than 80% of the total, was made up of artisans, blue-collar workers, labourers, domestic service workers and marginalised groups. The middle and popular classes suffered the crisis at the end of the 1800s and beginning of the 1900s, caused by the greater economic competition due to the arrival of agricultural products from outside of Europe within a context of growing economic globalisation. By the beginning of the twentieth century, the living conditions of the proletariat had deteriorated considerably (Parejo 1987a, 343–4).

Within the Andalusian context, Antequera was unique in the nineteenth century, particularly compared to the majority of the medium towns with a similar demographic size. Unlike these, it was not a strictly agricultural town forming part of a structure made up of large urban centres that together constituted districts with an agriculture able to market part of their surpluses. Neither was it an industrial town, even though, as we shall see, it underwent a successful process of industrialisation from 1833 and its industrial sector employed more than a 35% of is active population in the middle of this century. Antequera shared features of both models and had its own third characteristic derived from its privileged geographical location in the centre of Andalusia: a "natural passage" between the north and the south and the east and west of the region, and relatively close to Malaga and Seville, which, due to their ports, became distribution hubs for goods. In this way, Antequera became a space of transit, and, according to Professor Antonio Parejo, had "a certain level of cosmopolitanism", removed from "the profoundly rural and provincial mentality that characterised other towns of Andalusia". This was also due to the arrival of people from other places, particularly from Germany, the Basque Country, Galicia (Spain), and, most of all La Rioja (Spain) and France. Another salient feature of Antequera were the ties that its elite classes had with the political power in the capital, where some of its members had a prominent role, such as the Aguilares, led by Manuel María de Aguilar y de la Puerta, who founded the Democratic Party, or, among the conservative lines, Francisco de Paula Romero Robledo, who, together with Antonio Cánovas del Castillo, was the artifice of the political regime of the Restoration period (Parejo 1987a, 364–5; 1992, 18; Gómez and Parejo 2009, 11).

Within this uniqueness, during the 1800s Antequera prospered because it carried out a successful industrialisation process, based on "mechanisation and factory concentration", using mainly wool as a raw material. In the second third of the 1800s, it was the only medium-sized town of Andalusia (between 15,000 and 40,000 inhabitants) that was able to implement an industrialisation process based on an artisan tradition, which survived the political chaos and the economic disarticulation of the beginning of the century. In 1861, it held the third and fifteenth positions in the ranking of Andalusian and Spanish industrial towns, respectively. The wool textile activity entered a period of recession in the 1870s, which also affected other producing centres outside of Catalonia. This recession was a result of the technological delay and the competition from textiles of the same fibre and cotton from other parts of Spain and abroad. In 1887, Antequera "did not form part of the group of industrialised Spanish towns". At that time, the textile sector began a period of partial renovation, based on new technology and the substitution of the production of cloths for blankets. Other industrial sectors experienced different trajectories. For example, while the food and metallurgical sectors expanded at the end of the nineteenth century, the tanning sector was not modernised (Parejo 1987a, 225–237 and 333–6; Parejo 1987b, 258–9; Gómez and Parejo 2009, 16 and 21).

In the period of study, the demographic and urban planning trajectories of Antequera reveal the limitations of its economic growth model. The major difficulties experienced by Spain in the first two decades of the 1800s, shared by Antequera, explain that in 1825, with 15,836 inhabitants, it only had 380 more than in 1787. Subsequently, its population did not stop growing until the middle of the century: in 1838 it had 27,340 inhabitants, 21,675 in 1848 and 27,340 in 1857. From then, its population stagnated and began to show a decreasing trend, with 25,000 inhabitants between 1860 and 1895. At the beginning of the twentieth century, it increased again to reach 31,609 in 1900. In the first third of the 1900s, it remained stable with slight fluctuations: 32,363 inhabitants in 1910, 31,526 in 1920 and 32,974 in 1930, which reveals the existence of an unfavourable socio-economic context. If we consider only the urban population, the number of inhabitants began to decrease from at least 1887. This meant that the rate of urbanisation of the municipality, that is, the percentage of the population censored in the urban centre with respect to the total, decreased from 73.8% in 1887 to 65.5% in 1910 and to 57% in 1930. In these early decades, the evolution of the birth and death rates reflected a demography typical of the Former Regime, exceeding 30% and 25%, respectively. However, this affirmation warrants some qualification. The excess death crises were less recurrent and were caused by epidemics, such as that of cholera in 1835 and 1885 and the flu epidemic of 1918. On the other hand, the agricultural crises hardly affected the death rates (Parejo 1987a, 295; 1992, 20–1; Gómez and Parejo 2009, 12 and 17; Heredia-Flores 2011, 128).

At the beginning of the twentieth century, the major deficiencies exhibited by Antequera were attributable to different causes. Primarily, they were due to the afore-mentioned increase in social differences and also to the fact that the industrial activities followed disparate paths. There were those, as previously explained, which experienced enormous difficulties, while others prospered. There was a tendency towards industrial diversification with the particular development of the agro-industry, mostly the production of oil, sugar and beets. The growth of this industry can be related to the primary sector which increased is productivity and to the multiplication of the agricultural colonies together with the initiative of the large landowners and leaseholders who modernised their farms. This growing economic diversification did not contribute to reducing the social inequalities and had its correlate in the aforementioned demographic stagnation during the first three decades of the century.

3 Transformations in the Urban Layout and Public Services

The demographic evolution of Antequera can be linked to that of its urban layout, which, in turn, can be linked to that of its public services. During the Modern Age, the town had been shaped around the erection of several convents, which gave rise to its urban structure. The ecclesiastical disentitlement which began in 1836 implied the sale of a large number of properties of the Catholic Church distributed throughout the urban centre, the disappearance of one convent and the change in use of all of the others to residential, industrial and commercial purposes. In this way, Antequera accommodated is demographic growth in the central decades of the 1800s by increasing its density instead of its area.

In Estepa Street, the town's main street, the first pavements date back to 1838. It was not until the mandate of the mayor Francisco Guerrero Muñoz (1838–1906), that an ambitious reform programme was carried out, beginning in 1838, comprising the widening of the pavements, the paving of many roads, the remodelling and extension of the gardens and the renovation of the entrances into the town (Parejo 1987a, 301–2; Heredia-Flores 2011, 148–50 and 152).

The public lighting service also prospered during this mayor's term of office. From 1845, even though the oil lamps were installed in very few streets, the attempts to establish gas lighting failed. The oil lamps were replaced with petroleum lamps in 1880. Antequera was one of the first Andalusian towns to have electric lighting in its streets on a regular basis from 1892. It was implemented by *José Bellido y Compañía*, a local company established shortly after José Bellido obtained the concession to provide the service from the local government. Through a contract signed with the local government, the electricity company obtained the monopoly for the supply of public and private electricity for 30 years and committed to lighting the streets with 435 lamps. Electric lighting began to be installed on 13 November 1892 and by 1 December the whole network was operating. In the same month, the private lighting service had 136 clients. In 1893, the concessionaire began a project to reform and extend the public lighting services and in exchange obtained the extension of the contract for a further 10 years, that is, 40 years in total. In June 1897, the electricity company began another project to extend the services which would affect the main streets of the town and complement that of the "endpoint neighbourhoods". In around 1896, the business of *José Bellido y Compañía* was transferred to *Catalina y Simons Drocems*. At the beginning of the 1900s, new electricity companies emerged, such as *Felipe Herrero, Bernardo Bouderé y Sobrinos* (founded in 1902 by Bernardo Bouderé Bordenave and his nephews Bernardo and José Laude Bouderé, a family also dedicated to industry and banking) and the Carreira family (formed by large landowners who were also engaged in the production of oil, flour, wine, etc., led by José Carreira Gallardo). At that time, the latter obtained the public lighting contract, also with the electrical equipment of the Drocems. In 1910, the Bouderés provided the service to 600 private individuals and the Carreira family to 500. Two years later, the Bouderés dispossessed the Carreira family of the public lighting concession (Fernández-Paradas, 2006, 2010). In 1922, the *Cooperativa Eléctrica Antequerana*, promoted by the Luna family, took over the management of the network that supplied electricity to the town from the Chorro reservoir, controlled by *Hidroeléctrica del Chorro* and obtained the contract for public lighting. The new concessionaire also made a decisive commitment to expand the private line of the business (Parejo 1992, 27).

We will now contemplate the introduction of other important services. In 1804, the Magistrate Diego Sanz de Melgarejo began the initial procedures to construct the cemetery. The project was resumed in 1820 by the mayor Luis María Parejo Obregón y Rojas. Its construction began in 1829 and in 1840 it was operating on land located to the west of the town's periphery. It grew over the following decades and was expanded in 1909 (Flores Aragón 2019). Similarly, Antequera had a permanent telegraph station from fairly early on, from 1862, with a short delay of between five and seven years with respect to large cities such as Malaga, Barcelona and Madrid. The construction of the food market suffered a longer delay. Although there were projects to build it from 1866, the works finished in 1885. Its current location is in the historical quarter of the town, adjacent to the Church of San Francisco. As a result of the application of the Law of 25 October 1820 on the suppression of the monasteries and the regulatory reform, the Hospital de San Juan de Dios, located in Estepa street, was municipalised and was fully operating in November of that year (Escalante 2008). From 1856, a religious community of San Vicente de Paúl was responsible for caring for the sick (Heredia-Flores 2011, 154 and 156). There was a considerable delay in the arrival of the telephone. Although the operations to bring it to the town began in 1915, the concession of the service was granted in 1920 and it began operating on 19 February 1922 (Parejo 1992, 28).

Therefore, in general terms, the development of the public services was slow, which was due not so much to the disinterest of the municipal councillors of Antequera, but to the chronic lack of economic resources of the Spanish local governments. The origin of the problem resided in the centralised finance model created in 1845, which was maintained until 1924. This left little margin for the municipal corporations to decide on the type of administration that was most appropriate for their needs and prevented them from implementing municipal taxes. On a local level, the reform of 1845 was based on a direct tax, called the "Contribución Territorial", which did not generate the expected resources, due to the concealment of assets by some of the taxpayers who could not be identified given the lack of a register of urban and rustic wealth which would have obliged the richest segments to pay more. An attempt was made to counteract the insufficient resources with the creation of the consumer tax, which affected certain products, some of which were staple goods. This was not fair, as it affected the working classes more, who were obliged to spend a greater proportion of their income on food.

The transition from the Old Regime to a liberal society led to the local revenue authorities losing income from personal property, which was one of their largest sources of income. This was due to the disentitlement regulations, particularly Madoz's Law of Civil Disentitlement of 1855. As a result, the state exchanged the expropriated assets for debt securities and in the case of Antequera, for shares of the Cordoba railway company, although their value and the revenue that they generated turned out to be a ruinous business. This explains why the revenue collected from private property assets fell from representing 23.6% of municipal assets in 1885 to 4.5% in 1929. The result was a growing municipal debt, which, by 1885 represented 44.5% of the local government's budget.

Despite this, the local governments continued to be responsible for the same assistance and social obligations that they had traditionally held, such as prisons, hospitals, schools, etc., to which new functions were added. To compensate the growing imbalance between income and expenditure, the State allowed the municipalities to increase consumer taxes and to replenish with national taxes. However, this was not enough, and the result was the insufficient investment in public services (Parejo 1992, 35–8; Simón and Matés-Barco 2010, 31–4). This helps us to understand why until the 1920s the municipal corporations only directly managed those services which their feeble finances could cover and which did not require large investments, such as abattoirs, markets, cemeteries, etc. On the contrary, those services requiring large amounts of capital, such as the supply of water, gas and electricity were left in the hands of private companies (Núñez 1996, 402–3; Matés-Barco 2014). This also explains the different chronology in the development of the network services that required large investments. In the mid-1800s, the entrepreneurs had difficulties to raise considerable amounts of capital. During this period, the entrepreneurs in the gas industry were the only agents who were responsible for this service, although with limitations, given that in Andalusia and Spain this product was mostly supplied to the large and medium-sized cities. The other services, such as electrical lighting and transport were developed later, almost at the same time as the water services (Matés-Barco 2016, 23).

It was not until some time later, when the *Union Patriótica*, the party of Miguel Primo de Rivera's dictatorship (1923–1930), came into power in Antequera that, within the context of its public works promotion policy, a project of modernising public services was implemented. Although the project was not fully completed, the main streets of Antequera were paved, the gardened areas were extended and, as we shall see in the following section, the water supply and sewage networks were improved (Parejo 1992, 39).

4 The Slow Progress in Water Supply Infrastructures

The classical water system was in operation during the Modern Age. It had five main characteristics: technical limitations related to the unfeasibility of reaching all urban buildings and the linear nature of the aqueducts; the various types of supplies, with a differentiation between individual types (tanks and domestic wells) and collective types (aqueducts and irrigation ditches); the low level of quality control of the water; the low consumption of this resource per inhabitant, at around 5–10 litres per capita; and, finally, this system formed part of a model in which the use of water for irrigation was dominant with respect to the other uses (Matés-Barco 1999, 38).

From the mid-1800s, there was an increase in the need for this resource, due to the spatial and demographic growth of the cities and the emergence of new needs for street cleaning, industry, domestic consumption, and hygiene. At the same time, the demand for agricultural irrigation also grew (Larrinaga and Matés-Barco 2011). The amount of water was not only relevant for supply, but also because it was used to eliminate waste. The cholera and typhoid epidemics created an urgency to resolve "the problem of evacuating wastewater". The greater diversity in demand gave rise to pressure to ensure a domestic and continued supply. Hence, the legislators indicated a minimum consumption per person. Furthermore, the hygienists were successful in raising awareness about the need for cleanliness, promoting, for example, the need for street cleaning and the installation of bathrooms in homes and a new concern emerged regarding laundry. This group associated the use of water with improved health, and thanks to its growing influence, water treatment systems began to proliferate. Therefore, a new concept of the water supply service began to emerge. For this to become a reality it was necessary to replace the former system with the new one. This differed from the old system as it sought to provide a domestic and high-pressure supply, it was to improve potability through the use of chlorination and filters and it was to be regularly distributed throughout the whole town. In the modern system, there was a greater involvement of social agents (including the local governments) in resolving water supply problems with collective responses (Matés-Barco 1999, 44, 46 and 51; 2001, 138; 2016, 27–9 and 33; Novo 2002, 315–6).

From the beginning of the nineteenth century, legislation was approved to enable the municipalities to implement measures to organise the system so as to guarantee the supply and quality of water. This was the case of the Instrucción Municipal of 1813. Law 3 of 1923 approved the Instruction for the economic governance of the provinces, in force from 1836, which clarified the municipal nature of the supply of water and granted the local governments the power to enforce the safety of the resource, establishing that "they will also address the drying up of the lagoons or reservoirs and ensure the flow of stagnated or unsafe water". Furthermore, it established that the municipalities were responsible for the conservation and cleaning of public fountains and ensuring sufficient water for people and cattle. Similarly, it is worth highlighting that they were also responsible for "works for public use and adornment" and those of "utility or comfort of the neighbourhood". This included the water supply. The Municipal Law of 1877 confirmed the obligation of the local governments to provide this service. Furthermore, it specified its responsibilities, mentioning the supply of water, public wash houses, etc. The Public Works Law of the same year established that the local governments were responsible for supplying water. Two years later, the Water Law granted them the power to establish "regulations for the regime and distribution of water to their populations". The Hydraulic Works Law of 1911 indicated the responsibilities of the local governments for domestic supply. Two royal decrees of March 1914 and June 1925, respectively, regulated the support given by the State to the municipalities for the supply of drinking water, contemplating the concession of subsidies to undertake "water outlet, catchment and piping" works. Finally, the Municipal Statute of 1924, to which we will refer later, was completed with the Regulations for Works, Services and Municipal Assets, which established that the municipalities were responsible for resolving the applications for concession which could not be granted under a monopoly regime (Matés-Barco 2017, 44–8).

The supply of drinking water in Antequera from the Modern Age to the present day has always been managed by the local government. This is a unique case, given that, particularly from the 1800s, it was usually the case in Spain for the public service to be owned by the local governments but managed by private companies. This was due to the lack of economic resources of the municipalities and the fact that the modern water supply system required large investments for its implementation and maintenance. Furthermore, similarly to other network services, the supply of water could not be fragmented and required a complex industrial and commercial organisation involving qualified personnel which the Spanish local governments lacked when the technical network services began to be implemented (Simón and Matés-Barco 2010, 32, Matés-Barco 2016, 19). This is why the case of Antequera was different.

The supply of drinking water has two unique features. The first, which has already been mentioned, is that it requires large investments in assets which cannot be engaged in alternative uses. The other is that water is a necessity good which should be of a good quality, universal with a continuous supply, which implies sanitary control (Matés-Barco 2016, 15).

At the beginning of the nineteenth century, the town's water supply came from the sources of the Magdalena and Villa rivers, both located in a mountainous area and a distance of around 5.5 kilometres from the urban centre, the former to the west and the latter to the north. In around 1840, they supplied 11 public fountains and 31 individuals. There was a third river source, that of the Duranguillo, which supplied three public fountains and two individuals. At that time, only the wealthiest families had water in their homes; for the rest there were two options: either having a well in the home or recurring to public fountains (Heredia-Flores 2011, 151).

In the 1800s, the first project designed to improve the service overall was proposed in 1818 by the *Junta de Propios* created that same year. The reason behind it was the low amount of drinking water available to the municipality which had a quality that left much to be desired, being described as "dirty, filthy and degenerated". It was transferred from the Villa spring by an infrastructure with serious shortcomings, as the water was uncovered for a large part of the trajectory and the pipes were old and poorly built and filled with refuse. At that time, different possibilities were considered. There were those in favour of building a new aqueduct from the source of the Villa river, those who supported the continuation of the existing aqueduct, others who proposed the renovation of the Magdalena aqueduct or those who suggested a new course across the Duranguillo meadows.

Finally, the project comprised the building of a new aqueduct[2] which transferred the waters from the Magdalena spring, across the Duranguillo meadows, with a cost of 353,580 reales. The Royal Council of Castile ordered its execution on 23 March 1819.

A report issued by the Water Commission of the local government of 1838 defended the decision of the *Junta de Propios* to build a new aqueduct, arguing that the existing one passed through unstable ground with steep slopes. Therefore, if it had been decided to continue with the existing aqueduct, the investment would have been larger than that required for the new aqueduct. The report also explained that the work had not been finished because when the project was proposed, the revenue forecast with which to finance it seemed sufficient, taking into account the rustic heritage, the festival tax, the income from bull fights, the power to grant water straws and the dues from the large taxpayers. However, the reduction of capital resources and the changes in the government and local government negatively affected the progress of the works (Escalante 2008, 24[3]).

In September 1850, José Moreno Burgos elaborated a report informing that this work had not been executed. The report referred to the option of extending the aqueduct of the source of the Villa river or of building a new one with a different path in order to bring the water to Portichielo, located at the highest point of the town. He ruled out extending the latter due to its bad state of repair as a result of the poor quality of the land across which it ran and to the

2 For more on this project, see Archivo Histórico Municipal de Antequera (AHMA), Fondo Municipal (FM), Aguas y Riegos (AR), *Informe de la Comisión de Aguas sobre la conclusión del acueducto de La Magdalena*, Granada, Imprenta de Benavides, 1838.

3 AHMA, FM, AR, *Informe de la Comisión de Aguas sobre la conclusión del acueducto de La Magdalena*, Granada, Imprenta de Benavides, 1858.

fact that its layout prevented water from being supplied to half of the population. He also considered that building a new aqueduct was not feasible as the cost was too high and because farmers and industries would be deprived of the water that it had been supplying to them.

Moreno Burgos said that when the decision was made to build the new aqueduct from the Magdalena, the difficulties of the layout were not taken into account and the works were commissioned to the master builder José de Reyna, who did not consider the characteristics of the terrain. A total of 300,000 pesos were invested in its execution and the work could not be carried out. A proposal was made to renovate the existing aqueduct from the Magdalena as half of it was considered to be in a good condition. This option was based on the budget elaborated by Francisco de Torres, a master builder of the town, who established the cost of the works at 155,464 reales, which would be covered by the sale of the waters from the sources of the Magdalena and Villa rivers to individuals, for which the local government would build the pipes in the urban centre.[4]

In the mid-1800s, the use of water was a source of conflict as animal and human consumption, farmers, and industries, such as textile manufacturers and even flour mills and tanneries all competed to access the resource. In the nineteenth century, the water from the river Villa was the only source of energy used by the textile industries. The mills intensified the exploitation of the river by increasing the number of water wheels. They bought old mills and other buildings situated on the banks of this river in order to have a minimum infrastructure for the industrial use of the water and, most of all, so as to obtain its extraction rights. Therefore, different mills and wash houses were converted into factories or shared their primary use with textile production. The cloth manufacturers increased the volume of inlet pipes as the mechanisation of their factories intensified. Furthermore, they made new illegal courses and channels which gave them the theoretical right as owners of the properties that they had purchased. This gave rise to conflicts with the tanners, millers and farmers over the use of water from this flow, which was also irregular due to the dry season. This limitation prevented the factories from operating throughout the year. In 1860, they operated for an average of 194 days, although not throughout the whole day so that irrigators and other usufructuaries could make use of the water. The fact that the municipal by-laws of 1531, which did not contemplate the regulation of the common exploitation of the river, were still in force did not help to resolve the conflicts that were particularly intense

4 AHMA, FM, AR, *Informe de José Moreno Burgos*, 21 September 1850.

in the 1840s and prevailed until July 1854, when a Water Agreement was passed (Parejo 1987a, 328–30).

This agreement sought to establish a fair distribution of the time for using the water among the different usufructuaries, a fixed diameter of the pipes and the cut-off of the flow when water was scarce. Shortly afterwards, it was appealed by José Ramos Gaitero and "other industrialists and neighbours". Their arguments were mainly based on the fact that the agreement was made without consulting the overseeing mayor or the municipal commission and based on the vote in favour of one councillor and the mayor interested in its approval, so there was a clear conflict of interests. Furthermore, they argued that "the local government is not empowered to intervene in this matter", and the consent of the owners affected had not been sought and its "clandestine" use had not been addressed. The appeal was not successful, as we know that the agreement came into force in 1924.[5]

Eleven years after the 1854 Agreement, in 1865, a new important change was introduced. The local government, with Antonio López Gamarra y Nogueira as mayor, launched a public tender for the works to construct a new piping system from the source of the Magdalena river, based on the project of the engineer Deffarque. Two proposals were submitted for the tender held on 10 February 1866. One by Félix Martínez Rubio for an amount of 1,323,700 reales and one by Augusto Ruiz de la Herranz for 1,323,000 reales. The contract was awarded to the latter. The documentation consulted indicates that the local council paid for the works with the sale of new water straws to individuals and with the securities that it had in the Cordoba to Malaga rail company. In February 1867, the contractor informed the Municipal Corporation that the works had been completed.

This infrastructure was in operation until the beginning of the 1920s. In 1920, the town had hardly changed its appearance since the 1880s, except for the public lighting. In January 1921, the pages of the weekly local newspaper El Sol de Antequera, criticised the deterioration of the public services in the following terms:

> [...] this urgent and necessary transfer of waters has barely begun and should be reaching its end, if not suspended; the sewage and paving, which are publicly and daily condemned by the local residents, the reform of the abattoir, the repair of all the public buildings which are in danger of ruin due to their neglected state, in which the essential repairs

---

5 AHMA, FM, AR, *Copia del Convenio de 1854 de las Aguas del Río de la Villa.*

> for the conservation have not even been carried out; the termination of the cemetery works and its sanitation....not even one more cart for public street cleaning [...]. (Parejo 1992, 39)

This criticism reflected the demands of the wealthier households, including those of an emerging middle class who called for an improvement in services such as street lighting, drinking water, the telephone system, etc. (Otero 2020; Fernández-Paradas 2020). With respect to the water supply, there was an awareness that it was necessary to treat the water in order to prevent the propagation of diseases, particularly epidemics related to the poor state of the pipe networks that distributed it. To this, we can add the spread of better personal hygiene habits among the wealthier segments, who began to incorporate bathrooms in their homes and a more frequent laundering of clothes. Undoubtedly, this required a greater consumption of water per inhabitant.

The dictatorship of Primo de Rivera, in its quest to improve the operation and management of local governments, promoted the improvement of municipal public services. The Municipal Statute of 1924 gave them more autonomy to take on the responsibility for constructing infrastructures and services that could be funded through the transfer of charges and taxes. The reform of the water supply proposed by the local government of Antequera in 1927 should be placed within this favourable context. It responded to the new demands of the Municipal Statute of 1924 which obliged the municipalities to ensure the potability of the water and an amount of 200 litres per person per day (Matés-Barco 2016, 24).

This reform was part of a more ambitious plan which the local government commissioned to the Malaga-born Tomás Brioso Raggio (1885–1946)[6] in 1927. In addition to the water supply, it also contemplated paving, an urban

6 Brioso was educated in Madrid, where he obtained a degree in Exact Sciences from the Universidad Central in 1904 and an engineering degree from the Escuela Especial de Ingenieros de Caminos, Canales y Puertos. He developed his professional career in the public and private sectors, focusing on infrastructures and services: rail, irrigation, electricity, water supply, etc. Furthermore, he was a councillor of Malaga's local government during the dictatorship of Primo de Rivera. From this position, he promoted the Plan de Grandes Reformas y Mejoras de Málaga (Plan of Major Reforms and Improvements of Malaga) of 1924 and the Proyecto de Ensanche (Enlargement Project) of 1929. In 1930, after he had left the local government, he directed the works of the Chorro reservoir and the irrigation channels in the surroundings of the Guadalhorce river. With a conservative ideology, at the beginning of 1937, he was appointed Chief of Public Works of Huelva until October of that same year when he became the director engineer of the Port of Malaga. After the end of the Civil War, he had the same position in the port of Almeria. In the summer of 1944, he was promoted to advisory inspector of the Civil Engineering Corps (Heredia-Flores 2018; 2020, 261–7).

underpass, a perimeter park, the improvement of the educational supply and a ring road (Campos 2011 and 2013; Heredia-Flores 2020). This plan was feasible thanks to a substantial increase in the amount of the ordinary budgets, totalling 798,300, 1,137.000 and 1,698,000 pesetas, respectively in 1923, 1929 and 1935. In short, more income was dedicated to spending on public services. This item represented between 62% and 70.8% of the municipal budget in the 1920s and 1930s (Parejo 1992, 37). The triumph of the coup of Miguel Primo de Rivera in September 1923 in Antequera had its correlate in the renovation of the local power, which was led, during the dictatorship, by the greatest exponents of the local oligarchy group. These were Carlos Moreno Fernández de Rodas and José de Rojas y Arrese-Rojas, who alternated in the mayorship and León Serrailler Dromces, who directed the town's Unión Patriótica, the party promoted by the dictator.

With respect to the subject of this chapter, Brioso's plan, called the Sanitation and Improvement Projects of Antequera, was made up of different documents, including a report issued in April 1927 in which the engineer explained that his objective was to provide the city with a sufficiently abundant amount of water in good hygienic conditions. He affirmed that the two springs of the Magdalena and the Villa rivers had shortcomings. The first due to its scarce flow of around 20 litres per inhabitant per day, a very low figure. The second was more copious, but its waters were not apt for human consumption as troughs and washing areas polluted them. Moreover, it was limited by easements and exploitations located before and after the outlet that supplied Antequera. These deficiencies also affected the pipe network of the town which was laid with no predefined plan and was limited to the main streets. The scarcity of water in these pipes meant that they were not adequately cleaned.

In order to mitigate these shortcomings, Brioso planned to exploit to the maximum the capacity of the spring and the Magdalena aqueduct at around 15 litres per second and transport 20 litres per second from that of la Villa to Antequera in optimum conditions. Together, they guaranteed a total contribution of 55 litres per second, that is, approximately 160 litres per day per inhabitant, which reduced by half in the summer. In order to make this a reality, he considered compensating the beneficiaries of La Villa spring and constructing pipes to transfer the water to all of the homes, which would be sold to individuals based on the use of meters acquired by the local government and subject to a tariff. Furthermore, wastewater treatment plants and a regulating tank were to be constructed. The objective was to build a modern sanitation network with a length of 17,402 metres. He opted for a Waring model sewage system which was cheaper and easier to maintain, made up of circular pipes and an ovoid profile for the largest collectors, able to receive wastewater and rainwater. This was complemented with a network of automatic discharge chambers designed for cleaning the pipes and drains to receive rainwater.

The total cost of the Plan was established at 4,144,405.78 pesetas, with 1,204,635.40 and 879,096.70 pesetas assigned, respectively, to the water supply and sewage works, that is, 2,083,732.1 pesetas, 50.3% of the total amount of the investment. The water supply works would be financed by the local government (25%) and the beneficiaries (75%). For the sewage system, there was a 50% division.[7]

The processing of the extraordinary budget to finance the Plan was delayed due to changes in the mayorship, as José de Rojas y Arrese-Rojas was replaced in June 1928 by Carlos Moreno Fernández de Rodas. Shortly afterwards, the definitive budget was approved by the Municipal Corporation. This slightly reduced the forecast cost of the works referred to in this chapter, which amounted to 1,984,406.30 pesetas, of which those of the water supply totalled 1,147,271.80 pesetas and those of the sewage system 837.134,50 pesetas, 39.3% of the total budget.[8]

A commission was immediately created responsible for obtaining the loan, constituted by the mayor and councillors Juan Rodríguez Díaz and José de Rojas y Arrese-Rojas, who successfully negotiated a loan with the Banco de Crédito Local which the municipality took out in January 1929. Shortly afterwards, the contract for the sewage and water supply works was awarded to Esteban Pinilla Aranda associated with Santos Atienza. He made a bid for 671,790 pesetas for the water supply works and 525,339 for the sewage system, which were around 35% lower than the amounts established in the tender. That said, the execution of the works was delayed as in March of that year the Central Commission of Local Sanitation elaborated a report that approved the water provision project but not that of the sewage system. This obliged Brioso to design an additional project that removed the terminal siphon and incorporated a hydraulic lock in the sewage grates, which increased the cost of this project to 797,445.06 pesetas. The new document was approved by the Central Commission in May 1929. By August of the same year the execution works were well underway.

The fall of the dictatorship at the beginning of 1930 meant that those who wielded the local power at the time of the coup in 1923 recovered their positions. The mayor, José García Carrera denounced the paralysis of the works and delays in the payments of almost 124,000 pesetas. In October 1932, the local government approved the reception of the sewage and water supply works, even though these works were not finished until well into the 1930s (Heredia-Flores 2020, 298–309).

7 AHMA, FM, Proyectos de Saneamiento y Mejoras de Antequera. 1927, box 3.042.

8 AHMA, FM, book 3.581.

5 Conclusions

Antequera was a unique town, particularly in the 1800s, as, taking advantage of its privileged geographical situation in the centre of Andalusia, it successfully developed industrialisation processes, mainly based on wool textile, which began to show signs of exhaustion from the 1870s. In the decades at the turn of the century, its economy began to diversify although this process did not foster the tempering of social inequalities. This circumstance and its demographic evolution, particularly in the first third of the twentieth century highlight its lack of dynamism. Within this context, from 1918, the local government of Antequera promoted different initiatives to improve the water supply infrastructures. This first project failed. It was not until 1847 that a second major project was successfully completed, which substantially improved the transfer of waters from the source of the Magdalena river. The next new initiative was not proposed for another 60 years. It was larger and sought to implement the modern water supply system that became one of the most relevant landmarks in the history of the modernisation of public services in Antequera. Finally, it is worth mentioning that during the whole period under analysis, the management of the service was in the hands of the local government, a situation that prevails today. This constitutes a unique feature with respect to the overall pattern on a regional and national level.

6 Sources

AHMA, FM, AR, Informe de la Comisión de Aguas sobre la conclusión del acueducto de La Magdalena, Granada, Imprenta de Benavides, 1838.
AHMA, FM, AR, Informe de José Moreno Burgos, 21 de septiembre de 1850.
AHMA, FM, AR, Copia del Convenio de 1854 de las Aguas del Río de la Villa.
AHMA, FM, AR, Informe de la Comisión de Aguas sobre la conclusión del acueducto de la Magdalena, Granada, Imprenta de Benavides, 1858.
AHMA, FM, Proyectos de Saneamiento y Mejoras de Antequera. Año 1927, Box 3.042.
AHMA, FM, book 3.581.

References

Campos, J. 2011. "La Antequera del Sexto Centenario (1910–2010). La larga gestación de la ciudad media andaluza." In *600 años de Historia*, edited by J. Escalante Jiménez, 255–193. Antequera: Ayuntamiento de Antequera.

Campos, J. 2013. "El Plan de Grandes Reformas Urbanas de Antequera (1928)". *La Crónica. Comarca de Antequera*, 2–8 November 2013, 24.

Escalante Jiménez, J. 2008. *El abastecimiento de agua en Antequera. El memorial de Yllanes Sancho de Toledo (1545)*. Antequera: Aguas del Torcal.

Fernández-Paradas, M. 2006. "La Casa Carreira: un ejemplo de la pequeña empresa de electricidad en Andalucía (1903–1935)." *Baética: Estudios de Historia Moderna y Contemporánea* 28 (2): 583–600.

Fernández-Paradas, M. 2010. "Los comienzos de la electricidad en Andalucía: el ejemplo de Antequera (1892–1912)." *Baética: Estudios de Historia Moderna y Contemporánea* 32, 507–25.

Fernández-Paradas, M. (ed.). 2020. *Dosier La ciudad moderna. Infraestructuras en red. Andalucía en la historia 68*: 6–38.

Flores Aragón, A. 2019. "Historia del cementerio de Antequera." *Revista de Estudios Antequeranos* 22: 255–309.

Gómez Moreno, M. L., and A. Parejo Barranco. 2009. *Una economía en transformación: Antequera entre los siglos XX y XXI*. El Ejido: Fundación Cajamar.

Heredia-Flores, V. M. 2011. "La Antequera del Quinto Centenario (1810–1910)." In *600 años de Historia*, edited by J. Escalante Jiménez, 123–91. Antequera: Ayuntamiento de Antequera.

Heredia-Flores, V. M. 2018. "El abastecimiento de agua en Málaga (1860–1930): de negocio privado a servicio público." In *Agua y servicios públicos en España y México*, edited by J. M. Matés-Barco, and J. J. P. Rojas Jiménez, 121–53. Jaen: Universidad de Jaén.

Heredia-Flores, V. M. 2020. "El ingeniero Tomás Brioso y el plan de grandes reformas de Antequera." *Revista de Estudios Antequeranos* 23–24: 253–314.

Larrinaga, C., and J. M. Matés-Barco. 2011. "La modernizzazione delle città spagnole: il servizio di approvvigionamento di acqua potabile (1870–1936)." *Memoria e Ricerca* 36: 29–43. https://doi.org/10.3280/MER2011-036003.

Matés-Barco, J. M. 1999. *La conquista del agua. Historia económica del abastecimiento urbano*. Jaen: Universidad de Jaén.

Matés-Barco, J. M. 2001. "El servicio de abastecimiento de agua potable: estado de la cuestión.". *TST. Transportes, Servicios y Telecomunicaciones* 1: 135–58.

Matés-Barco, J. M. 2014. "Las empresas concesionarias del servicio de abastecimiento de aguas potables en España (1840–1940)." *TST. Transportes, Servicios y Telecomunicaciones* 26: 36–7.

Matés-Barco, J. M. 2016. "La regulación del suministro de agua en España: siglos XIX y XX." *Revista de Historia Industrial* 61: 15-48.

Matés-Barco, J. M. 2017. "El servicio público de abastecimiento de agua en España (siglos XIX y XX). El proceso de acumulación de competencias de los ayuntamientos." *Revista Brasileira de História & Ciências Sociais-RBHCS* 9 (18): 36–57. https://doi.org/10.14295/rbhcs.v9i18.448.

Novo López, P. A. 2002. "El agua potable a domicilio y los residuos urbanos bailan un complejo pas a deux (1850–1930): una aproximación metodológica e historiográfica." *Historia Contemporánea* 24: 281–319.

Núñez, G. 1996. "Servicios urbanos colectivos en España durante la segunda industrialización: entre la empresa privada y la gestion pública." In *La empresa en la historia de España*, coordinated by P. Martín-Aceña Manrique, and F. Comín, 399-422. Madrid: Civitas.

Otero Carvajal, L. E. (ed.). 2020. *Las infraestructuras en la construcción de la ciudad moderna. España y México, 1850–1936*. Madrid: Catarata.

Parejo Barranco, A. 1987a. *Historia de Antequera*. Antequera: Obra Cultural de la Caja de Ahorros de Antequera.

Parejo Barranco, A. 1987b. *Industria dispersa e industrialización en Andalucía. El textil antequerano, 1750–1900*. Antequera: Ayuntamiento de Antequera and Universidad de Málaga.

Parejo Barranco, A. 1992. "Introducción." In *Antequera, memorias de una época. Cincuenta años de la vida de una ciudad a través de la fotografía, (1885–1935)*, edited by A. Parejo and J. Romero, 13–44. Antequera: Biblioteca Antequerana.

Simón, I., and J. M. Matés-Barco. 2010. "El abastecimiento de agua potable en México y España: Cambio institucional y aparición de las empresas privadas (1870–1930)." In *Agua, territorio y medio ambiente. Políticas públicas y participación ciudadana*, edited by J. R. Navarro García, J. Regalado Santillán and A. Tortolero Villaseñor, 21–48. Guadalajara: Universidad de Guadalajara and Seminario Permanente ATMA, CSIC.

CHAPTER 7

Foreign Companies, Public Intervention, and Ecological Crisis: Water in Seville, 1871–1959

Alberte Martínez-López

Abstract

This chapter seeks to analyse, from a business history perspective, the role played by British capital in the management of the water supply in Seville for almost seventy-five years through the company *Seville Waterworks*. Using primary business and public sources, it examines the origin of the company, its shareholding and organisational structure and business results, which were conditioned by demographics and rainfall, the different economic situations, and the public regulations, all within the framework of the evolution of the water sector in Spain and the British investment abroad.

Keywords

ecological crisis – foreign British investment – free standing company – municipalisation – water supply

1 Introduction

Although Seville was one of the principal Spanish cities, its economic situation in the final decades of the nineteenth century, and to a lesser extent in the first third of the twentieth century, was sluggish. Local investments were focused on property and trade (Posadas 1991). The traditional overcrowding, aggravated by demographic growth (Table 7.1) and the deficiency of the infrastructures led to serious public health problems with high rates of mortality, particularly in the working-class neighbourhoods (González 1984; Barrionuevo 2005). This prompted the implementation, in the second half of the nineteenth century, of the new urban services that symbolised modernity; gas (1854) (González 1981), the railway (1859), the telephone (1880), water (1883), trams (1887), and electricity (1894), in which foreign capital played a leading role (British in water and mule-drawn trams and German in electricity and electric trams).

 | DOI:10.1163/9789004679634_008

TABLE 7.1 Population of Seville and inter-census growth 1860–1960

Year	Population	Inter-census growth
1860	118,298	
1887	143,182	24,884
1900	148,315	5,133
1910	158,287	9,972
1920	205,529	47,242
1930	228,729	23,200
1940	312,123	83,394
1950	376,627	64,504
1960	442,300	65,673

SOURCE: POPULATION CENSUSES

However, these reformist measures clashed with the rent-seeking interests of the powerful *Liga de Propietarios de Fincas Rústicas* (Association of Rural Property Owners), founded in 1894, opposed to any investment in infrastructure, particularly in the case of water, which represented an increase in the fiscal pressure on property. This opposition was shared by the Chambers of Urban Property of other cities (Braojos, Parias, and Álvarez 1990; Calvo 1994).

Foreign capital played an important role in the implementation and management of the public urban services in Spain and other peripheral countries, due to the reticence of the Spanish capitalists to finance activities that involved high investments and required technological and network management knowledge. In the case of water, the technical restriction was lower, which explains the greater relative presence of local companies. Most of the capital in this activity was French, followed by British and Belgian capital (Costa 1981). The emergence of foreign investments in this sector took place in the last two decades of the nineteenth century, resulting from a convergence of the interests of banks, water management companies and pipe manufacturers (Matés-Barco 2002). From 1920, as in the case of other public services, the majority of foreign capital began to withdraw, particularly from the large cities. On a geographical level, the investments focused on medium and large-sized cities, such as Barcelona, Seville, Alicante, or Las Palmas, where the profit expectations were higher due to economies of scale and possibilities to expand the networks.

British foreign investment in the water sector occurred relatively late, in the period 1885–1914 (Stone 1999). It came about due to an exhaustion of the possibilities of expansion for the sector in the domestic market derived from the intense municipalisation process of the activity carried out in Great Britain in the last third of the nineteenth century (Millward 2007; Hassan 1998). Furthermore, investment in water services was secondary compared to the other urban infrastructures. In the Spanish case, British investment in the water supply was modest, both in absolute and relative terms. It was irregular over time and occupied an intermediate position among the British interests in Spanish urban infrastructure. In geographical terms, the British focused their initiatives on territories where they already had economic and strategic interests (mines, ports, winemaking), such as Andalusia (Seville, Cadiz, Huelva, Algeciras), Las Palmas, Alicante or Cartagena (Broder 1981).

2 Origin of the Water Services and of *Seville Waterworks*

The water in Seville had traditionally been obtained from the river Guadalquivir. There were two problems: its brackish nature due to the tides and human contamination. Therefore, it was decided to draw resources from upstream. This supply was limited to the poorest segments of the population, who also used wells, which were also often brackish and unsanitary. From 1172, the wealthiest classes were supplied by the Santa Lucía spring, in the municipality of Alcalá de Guadaira, via a 17-km long aqueduct called "Caños de Carmona". It re-used the Roman aqueduct and constituted the basic water supply system of the pre-industrial city. However, the lack of adequate maintenance led to an alarming reduction in the flow available from the beginning of the nineteenth century, aggravated by the growing contamination of the wells and river. Sanitation was even more precarious. The sewage network was insufficient, with frequent breakages and blockages (Collantes et al. 2011; Fernández 2011). The urban, demographic, and industrial growth resulting from industrialisation worsened these problems. In order to resolve them, in the mid-nineteenth century the aqueduct underwent renovation works, which finished in 1871, and new sources were added to improve the quality of the supply. These measures were insufficient. The city council, therefore, decided to tender the management of the city's water supply.

As was usually the case in the system of public services concessions in Spain, a series of projects presented by Spanish engineers and foreign companies (French) ended in failure. They were generally speculative in nature and had little business and financial solvency. Finally, in 1882, the city council approved the proposal presented by Georges Higgin Winfield, an engineer from London who worked in several British mining railway companies in Spain. The 99-year long concession established that the water for domestic use would be drawn from different springs of the Guadaira river basin, about 14 kms from Seville and 2.000 m^3 for irrigation would be obtained from the river Guadalquivir. The minimum volume of water to supply would be 100 litres per day per person. This amount was lower than the 150 litres recommended but exceeded the 29–43 litres that, depending on the season, the people of Seville were receiving at that time. The maximum price established was 50 cts/m^3 for individuals, 25 for industrial uses and municipal buildings and 15 for municipal irrigation. As was customary in these processes, in December 1882, Higgin communicated the transfer of the concession to the construction company James Easton and Shaw,[1] which, the following year, transferred it to *The Seville Waterworks Company Limited*, the former undertaking the building works, in exchange for £200,000 in shares and £100,000 in bonds. The service began operating in 1885. For the first five years it was managed by Easton & Anderson which paid £21,000 per year in rent to *Seville Waterworks* in exchange for the revenue from the water. Finally, in 1890, *Seville Waterworks* received the concession and installations for 4,961,304 ptas.

The Seville Waterworks Company Limited was founded in 1882 in London as a free standing company. This type of corporate form was frequently used by small British companies operating abroad (Wilkins 1998). Its principal feature was that it was an independent company, not forming part of a business group and its initial vocation was to operate in a foreign country. Another of its features, which was common to the British companies, was that it was financed principally through the capital market and not so much the banking entities as was the case in the German and Belgian companies. It was the oldest water management limited company with the largest capital in Andalusia (Matés-Barco 1997).

The founding capital amounted to 25,000 shares at £20, of which, in 1887, only just over half had been subscribed and paid up, 12,906 shares, that is, £258,120, with a maximum being reached in 1893 with £271,120. Bonds had been issued with a 6% interest for the value of £100,000. The number of shareholders was very high, which was frequent in the developed British capital market and in investments considered as being safe, such as those of public services. The immense majority of the shareholders and, particularly of the shares, were British, some of whom were residents in Seville and a small minority were Spaniards, all of whom lived in Seville. There was not a high concentration of capital, as the five largest shareholders only held 19.3% (Table 7.2).

1 In 1883, Higgin obtained the concession for the tram, which he transferred in 1887 to The Seville Tramways, which had no connection with *Seville Waterworks* (Braojos, Parias, and Álvarez 1990).

TABLE 7.2 Shareholders in 1891

	Number	Number of shares	Shares by shareholder
British	170 (91.4%)	13,164 (98.2%)	77.4
Spanish	16 (8.6%)	242 (1.8%)	15.1
Total	186 (100.0%)	13,406 (100.0%)	72.1
Main shareholders			
Lord Hillingdon		750	
C. W. Mills		500	
D. Evans		480	
A. H. Mills		450	
R. Easton		405	

SOURCE: PRO, *SEVILLE WATERWORKS* 1891

Its organisational structure was simple. There was a Board of Directors, based in London, formed by a chairman, secretary and board members and a manager in Seville, in charge of the operating and administrative personnel. The latter managed the day-to-day running of the company but had to consult any important decisions with the directors in London.

The Board of Directors and the management were characterised, overall, by their stability, as the majority of the members remained in their positions for a long period of time, sometimes being replaced by their sons, often only due to death. The chairman was usually one of the members of the board with the most experience. All of this favoured the continuity of the business strategies in a sector which, in any case, did not usually experience complications, given its continually rising demand and the absence of competition. This desire for stability was also exhibited in its financial relations, as its bank was always the same for the whole of its existence, namely *Glyn, Mills, Currie and Co.* Only in the early years, and most of all, at the end of the company's life was there greater instability among the directors. In the initial phase, due to the sudden disappearance of the only two Spanish members, most probably named to facilitate contact with the Spanish authorities. In the final phase, this instability was due to the internal tension of the moment.

The management and technical side was undertaken by the manager who was always an engineer and British. In addition, at least at certain times, there was a sub-manager and/or assistant manager. They were also engineers and assisted the manager or were responsible for occasional projects. The only presence of Spanish engineers that we have found was that of Joaquín R. Garay after the First World War as a sub-manager. The management of this company, therefore, was always in the hands of British engineers. This was contrary to other British companies in Spain in which a progressive allocation of responsibility was given to Spanish professionals, particularly from the beginning of the twentieth century.

3 From the Start of the Service to the Agreement of 1912 with the City Council, 1885–1912

In the early years, the evolution of the company was positive. The network expanded and in 1896 it began to supply the municipality of Alcalá de Guadaira. All of this led to an increase in the number of users[2] and the water sold (Figure 7.1). In 1891, the majority of consumption, 82.1% corresponded to free pipe users, that is, at a fixed charge, 11.8% to companies, 4.4% to public thoroughfares and fountains and only 1.7% to contracts with a meter. In monetary values, individual consumption grew from 78.3% in 1890 to 81.6% in 1895. Municipal consumption fell from 20.7% to 17.1% and port consumption grew from 0.8% to 1.1%.

Logically, revenues also increased, while costs did not rise in the same proportion due to the growing economies of scale. However, the profits

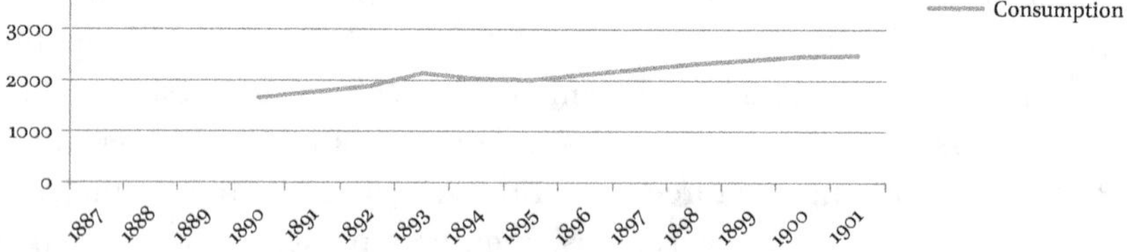

FIGURE 7.1 Water contracts and consumption, in thousands of m^3, 1887–1901
SOURCE: *ANNUAL REPORTS* 1887–1901

2 In 1887, the contracts represented 2.6% of the population, increasing to 5.4% in 1901. They were still modest figures, meaning that domestic water was only available to 20%–25% of the population.

expressed in pounds, although positive, were reduced by the negative evolution of the exchange rate. In any event, the economic and financial profitability (ROA and ROE) was the highest in the history of the company, enabling a regular and acceptable distribution of dividends at 3%–5%.

The increase in demand and the drought of several years forced the search for new water supplies and placed pressure on the costs of pumping due to the lowering of the water tables. In order to finance the works for the supply, installations and network expansion, the company issued new bonds, all at 5%: £20,000 in 1890, £10,600 in 1895–1897, £6,300 in 1896–1897. In order to reduce its financial burden, in 1891 it approved the reduction of the interest of the initial bonds from 6% to 5% (Figure 7.2).

The year 1897 marked the beginning of the problems for the company. The Cuban war caused a strong increase in costs, particularly that of coal, as well as the taxes and also a fall in the exchange rate (Figure 7.3). On the other hand, the long drought obliged the company to purchase water from the town council of the Santa Lucía spring for three years.

The forecast production in the concession was 20,000 m³, but due to the exhaustion of the initial springs, another three sources had to be used. Even so, they were still insufficient for the increasing demand, which obliged the company to sign a complementary contract with the city council in 1901 to use the "Caños de Carmona" in exchange for undertaking its maintenance and renovation. These waters were of a lower quality and accounted for 20% of the water supplied by the company. In 1912, the city council withdrew this use as the company failed to fulfil its obligations and began to manage it directly (Aparicio, 1990). In the stipulations of 1901, the city council committed to purchase from the company water from the Guadalquivir for irrigation and

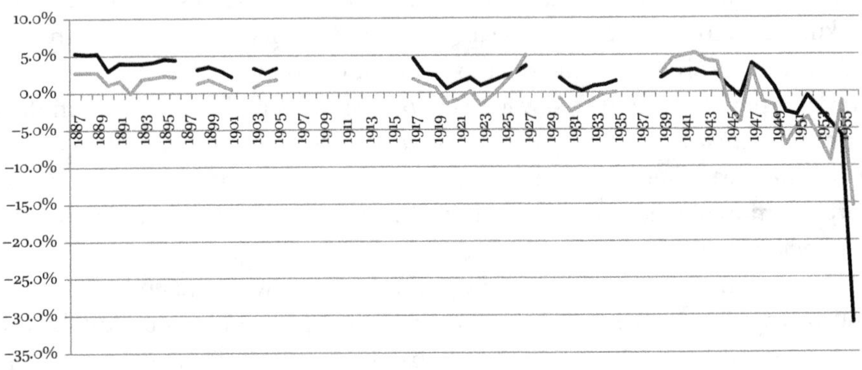

FIGURE 7.2 Economic and financial profitability, 1887–1956
SOURCE: OWN ELABORATION BASED ON *ANNUAL REPORTS* 1887–1901

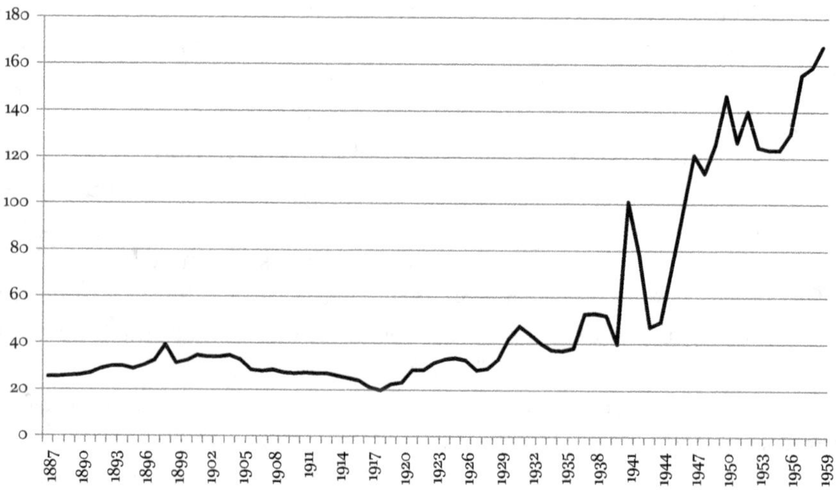

FIGURE 7.3 Exchange rate pta/pound, in pesetas per pound, 1887–1959
SOURCE: CARRERAS AND TAFUNELL 2005, 703–6

municipal services. The minimum amount was 731,000 m^3 per year at 10 cts/m^3 for irrigation and 15 cts for the rest. The rates of the new free pipe contracts for individuals were modified, increasing the number of groups, which ranged from 2.50 ptas per month to 25, depending on the rents. Industrial services would pay 25 cts/m^3. The municipal debts with the company would be paid in five years with a surcharge of 6% per year. The conflicts between the public service concessionaire companies and the local governments were commonplace and were usually caused by the accumulated debt, the rates, the quality of the service and the renewals of the concessions.

The increase in the available water also led to an increase in wastewater, making the already previous precarious situation unsustainable. In order to definitively resolve this situation, the city council commissioned the construction of a new and modern sewage network with a length of 150 kms, in 1899 with *Compañía Anónima de Saneamiento y Urbanización de Sevilla*, created that same year. Due to the delay in the construction works, sanitation was partially municipalised in 1920 and the new network began operating in 1926 (Aparicio 1990).

The situation of the company began to improve with the end of the war and the agreement with the city council in 1901. In order to undertake the new investments (wells, pumping stations), in 1901 the company approved the amortisation of the bonds in circulation and a new issue of £160,000 at 5%.

Throughout its long history, the manager's reports frequently referred to the level of rainfall, particularly in times of drought, as a highly relevant element

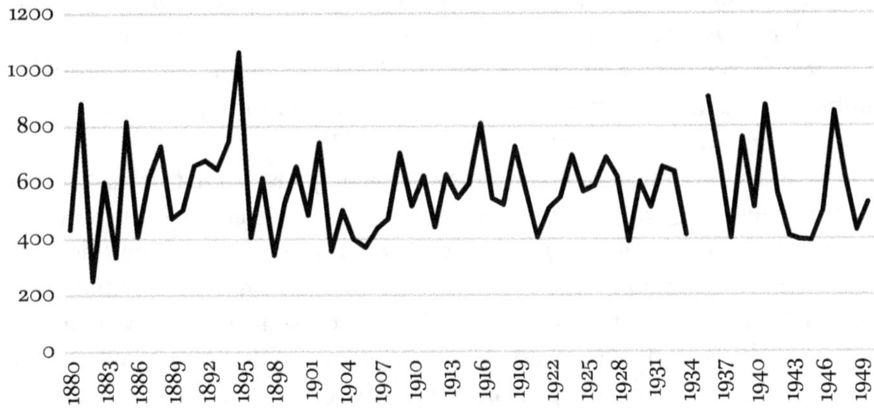

FIGURE 7.4 Rainfall in Seville, 1880–1950, in mm
Source: *Resumen de las observaciones meteorológicas efectuadas en las estaciones del Servicio Meteorológico Español durante el año... 1880–1950*

in the capacity of the company to provide the service. The Mediterranean climate of Seville did not help to replenish the aquifers. In the period 1880–1950, the annual average rainfall was only 522 mm, but the irregularity was highly pronounced, as the standard deviation was 156.5 mm and the coefficient of variation was 27.4%. Furthermore, there was a sharp fall in rainfall during the summer (Figure 7.4).

The overwhelming predominance of the free pipe contracts represented an incentive for abusive water consumption and aggravated the always threatening gap between supply and demand. Aware of this, the company began a campaign to install meters in homes.[3] In other cities, this measure was highly conflictive due to the opposition of the users and owners with the abandonment of the free pipe system due to its cheapness and lack of control (Calvo 1994).

4 From the 1912 Agreement to the Civil War, 1912–1936

Contrary to the demographic stagnation of the eighteenth and nineteenth centuries, the population grew significantly during the first third of the twentieth century due to the intra-provincial migratory movements, attracted by

3 On 31/3/1903, only 6.5% of contracts were with a meter, on 31/3/1905 this percentage had risen to 11%.

an incipient industrialisation and the Ibero-American Exhibition of 1929. This would have aggravated the traditional housing and sanitation problems. This meant that the preparation for the international exhibition had to include an ambitious urban remodelling which was coordinated with the hydraulic plans (supply, sanitation, and flood control) (Braojos, Parias, and Álvarez 1990).

The local tax office experienced two major phases. First, from the end of the nineteenth century until 1918 there was a stagnation in income and an imbalance due to growing costs (personnel, water supply, flood defence, Ibero-American Exhibition). Second, between 1919 and 1930, per capita income tripled thanks to the reform of the municipal tax system, which gave rise to an enormous increase in the investment capacity of the local government, also based on heavy indebtedness which, with the crisis of the 1930s, had to be partially written off by the government (Braojos, Parias, and Álvarez 1990).

Due to the growing mismatch between supply (reduction in the flow of the springs) and demand (demographic and industrial growth) and after laborious negotiations, a new agreement was reached between the company and the city council on 26 October 1912 to extend the supply. A double pipe and supply system was established: drinking water for human consumption taken from the Alcores (Alcalá de Guadaira) and filtered water from the Guadalquivir for irrigation and fire hydrants. The city council would have one million m^3 per year for 225,000 ptas. The debt of 24,000 ptas of the city council with the company would be paid in eight yearly instalments of 3,000 ptas.

The First World War complicated the application of this agreement and the very financial stability of the company. The maritime transport problems hindered and increased the cost of the import of materials (pipes, coal). Inflationary pressure, in terms of materials and wages, worsened in the post-war period, which made it very difficult to maintain a balance of the exploitation with the rates agreed in 1912. Therefore, the company negotiated a price increase with the city council. Despite the increase obtained in 1918, it was insufficient due to the continuous increases in costs.

Within this context, in 1922, a new agreement was signed with the city council to increase the volume of water taken from the river. The city council made a debt issue to finance the works. Meanwhile, the company began to exploit new wells and had to adapt the installations and extend and improve the distribution network. During this modernisation process of the installations, the electrification of the pumping stations began and was completed in 1927, generating a significant reduction in costs. Measurement instruments were installed which enabled any failure in the overall efficiency of the system to be located. The maintenance of the meters was reinforced and sterilisation with chlorine began in compliance with the new regulations. Despite the extensions and improvements, the reality was that the daily supply per inhabitant was practically the same as at the end of the nineteenth century and was still a long way off from the minimum of 100 litres stipulated in the concession.[4]

In parallel, the city council contemplated a possible municipalisation of the service, in line with the debate open in Spain from the second decade of the twentieth century, with a certain delay with respect to other European countries (principally, Great Britain, Germany and Italy) where the municipalisation of services had been a reality since the end of the nineteenth century. Also influential was the context of the mass withdrawal of foreign capital in local public services between 1918 and 1923. For this reason, the municipal technicians elaborated a report in 1925. They valued the installations of the company at 9.5 million pesetas. The network of the Guadaira was 150 kms and that of the Guadalquivir 100 kms. They considered that the financial situation of the company was critical. Municipalisation would oblige the city council to issue capital worth 12 million pesetas, install mandatory meters and increase the rates to 75 cts/m^3. The alternative in order to maintain the equilibrium of the exploitation would be to authorise an increase in the rates to the concessionaire or for the city council to assume the costs of extending the supply. Finally, the city council opted for a middle path: it municipalised only the supply of the water from the river in 1925.

In the 1930s, the situation worsened for the company due to a change in the economic situation and the new socio-political context. The economic failure of the Exhibition, coupled with the Depression led to a high level of bad debt, shared by other public services. This debt was mostly accrued by the users who received common water in the new urban development Ciudad Jardín, of the developer Nervión. The company bought the Ciudad Jardín network from Nervión in 1934 and progressively signed contracts for individual meters with these people (550 contracts). The city council could not make the payments for the water consumed or pay the interests of its bonds, the majority in the hands of the company.

To all of this we must add the wage increases derived from the labour disputes, the new charges and fines for not supplying the quantity stipulated in the concession, the strong fall in the exchange rate and the prolonged droughts. This latter factor reduced the water reserves, necessitating the digging of new and deeper wells in Clavinque. The company attempted to improve the efficiency of the system, intensifying the inspections of the meters to prevent leaks and illegal diversions, reviewing the long-term contracts for low-income homes

4 80.2 in 1890, 83.7 in 1927, although in the latter case the filtered water from the Guadalquivir managed by the city council would have to be added.

and establishing controls to prevent the entrance of refuse in the distribution network. It also requested an increase in the prices, although unsuccessfully.

On the other hand, and to prevent the loss of currencies and a greater deterioration in the exchange rate, the Spanish authorities blocked the outflow of capital. This prevented transfers in pounds of the business profits from Seville to London. In view of this critical situation, the company approved a three-year moratorium which was extended several times (until 30 September 1936) on the payment in cash of interests and the amortisation of the bond.

In 1933, the municipal technicians elaborated a report relating to the complementary water supply project. At that time, the city had three types of water service:

1. Those derived from the Caños de Carmona. This was an old, 12km-long aqueduct, half of which was in poor condition. It provided an average daily flow of 5,000 m^3. The waters were contaminated downstream from Alcalá de Guadaira. The pressure was insufficient for the water to reach the flats. The city council was responsible for the conservation and distribution.
2. *The Seville Waterworks*. Water collected by filtering galleries close to Alcalá de Guadaira or through wells in the Clavinque estate. Average daily flow of 15,000 m^3. Deficient and intermittent service. In general, the water did not reach the flats.
3. Municipal service. Filtered water from the Guadalquivir. Filtering installations for 25,000 m^3 per day. Highly variable chemical conditions depending on the seasons and river flow. These waters had to be replaced by the waters whose collection was being planned.

5 From the Civil War to the Municipalisation of the Service, 1936–1959

The Spanish Civil War and the subsequent World War hindered and increased the cost of obtaining energy and all types of materials. The autarchic and interventionist policy of the new regime altered the change process and working conditions. The exchange rate of the peseta was fixed by the government with different values depending on the purpose. The outflow of currency required government authorisation. In order to compensate the high inflation of the whole period, the paternalistic labour policy drove wage increases and different social securities. On the other hand, the intense migratory flows from the countryside to the city accelerated in the 1950s, leading to a persistent and significant increase in the demand for water.

Within this context of cost inflation (labour, electricity), pressure from the demand and strong state interventionism, the company had to act and repeatedly requested price increases. They were granted but only in small amounts and with long delays due to the governmental policy of controlling inflation. Therefore, they were insufficient. All of this deteriorated the financial position of the company, until it became clear that the best option for all parties would be the municipalisation of the service. The question now was the amount that should be paid by the city council for the early rescue of the concession. The situation of the local tax office did not allow it to undertake this type of spending. Therefore, the municipal authorities were protracting the negotiations, most probably because the economic suffocation of the company forced it to negotiate downwards. In the light of this impasse, the company recurred to the British embassy for it to mediate with the Spanish government so as to accelerate the decision-making with respect to the municipalisation of the service and to approve the increases in rates that were essential to prevent bankruptcy. As in the case of other British companies with difficulties in Spain (Martínez-López 2020), the embassy of the United Kingdom in Madrid took the issue very seriously and took action with the different Spanish ministries involved (Finance, Home Affairs, Foreign Affairs), expressing, with a measured and diplomatic tone, the points of view of the company. All of this was done after previously informing the Foreign Office in London and obtaining its acquiescence. This office also maintained close contact with the Board of Directors of the company. Below, we will analyse how this process developed.

The city council constructed the Algaba dam in 1937 to collect water from the river Rivera de Huelva, close to Seville, and not dependent on the Guadalquivir. The following year, it approved the project of the new water supply for which the Minilla reservoir was constructed. It became partially operative in 1946, enabling the storage of 20.000 Hm^3, together with certain sections of the piping channel to the city (González 1981). Due to the insufficiency of the municipal resources, the *Confederación Hidrográfica del Guadalquivir* intervened, which elaborated a Supply Plan in 1950. This Plan established a supply of 250 litres per day per person in the city. The works consisted in the finalisation of the Minilla reservoir, operative in 1957, the 42 km-long piping channel, the processing station and the headwater pond of 200,000 m^3 (Aparicio 1990).

From 1940, the company obtained authorisation to send, at the exchange rate and amounts fixed by the government, the profits to London. This enabled it to pay the interests of the debentures and the accumulated delays, although the moratorium continued to be extended. The distribution network was expanded to meet the growing demand and urban expansion to new neighbourhoods. However, this clashed with the difficulty to obtain pipes and other types of materials. In 1942, the government approved the replacement of the free pipe contracts with meters and a minimum consumption. The new flat rate would be 0.56 ptas/m^3, as opposed to the previous 0.32–0.50 pesetas. In 1945, all of the more than 7,000 free pipe contracts had been transferred to meters.

The water rates in Seville were frozen from 1918. In 1936, the company requested a rise and, after arduous negotiations, obtained an increase considered to be insufficient. In around 1938, the Seville water rates were the seventh cheapest of 35 Spanish cities. The civil and world wars postponed new appeals. At the end of 1944, the company renewed its request. The rate requested was 0.75 ptas/m^3. Immediately after the world war, the first appeal for help from the company to the British government was made with respect to this matter. From September 1945, interviews were held between the manager and Spanish authorities (mayor, Undersecretary for Industry), and the commercial attaché to the British embassy in Madrid (with the Undersecretary of Trade, Industry, Ministry of Industry and Trade, Ministry of Foreign Affairs) and much written correspondence between the embassy, the Foreign Office and, to a lesser extent, the company took place. The usual dynamics consisted in the company transmitting its request for help to the Foreign Office, which passed it on to the embassy in Madrid, which made a confidential report based on the information that it could gather and the actions it could take with mostly second-level officials of the ministries involved. Finally, the Foreign Office informed, after filtering the internal comments, the company of the actions taken. In the final stage of the negotiations, the Foreign Office consulted the Treasury.

The Spanish authorities were aware that the delay in the approval of increase in rates could be interpreted as a deliberate manoeuvre to financially suffocate a foreign company. Within a context of a desire for the new regime to break the international isolation, the Ministry of Industry maintained a diplomatic attitude with the British embassy, ensuring that there was no such discrimination, and that the decision was speeded up. At the end of 1945, the new rate of 0.75 ptas/m^3 was approved, increasing from the former 0.50 ptas. Further increases were subsequently obtained: 1.13 (1947), 1.22 (1949), 1.65 (1951). However, the company considered them to be insufficient and that they had been adopted too late. It alleged that while the wage increases and social norms had to be applied immediately, the authorisation of new rates suffered a very long delay, a period in which the company operated with losses.

The growing labour costs constituted a serious problem for the company. Between 1937 and 1949, the administrative staff in Seville had doubled and the cost had multiplied by five due to the increase in the workload (growth in the number of users, switching from free pipe to meter contracts) and wage and social improvements. Another significant problem was the progressive deterioration of the exchange rate, in response to which the company attempted to speed up the sending of remittances. The frequent power cuts, particularly during the summer drought, negatively affected the pumping stations and reduced the water supply. The losses in distribution continued to be high. A significant part was due to small leaks inside the homes, which did not compensate the replacement of meters. From 1950, considerable losses were incurred, and the company could not meet the payments of the interests of the bonds, The prolonged lack of profitability for the shareholders led to growing internal tension.

From 1951, the company was fully aware of the economic non-viability of continuing to manage the service under these conditions, particularly with the scarce and decreasing availability of water in the Clavinque basin. This was the structural cause of its problems, together with an expanding demand. Therefore, with the support of the embassy, it attempted to open negotiations with the city council and Spanish authorities for the early rescue of the concession and the temporary approval of a rate that would prevent its bankruptcy. The delicate financial situation of the British company was not an exception. The majority of the public services companies in Spain, particularly essential services such as water, had serious financial problems due mainly to the mismatch between the continuous cost inflation and the approval of clearly insufficient rates. This led to the municipalisation, with governmental financial support, of many local water concessions in the 1950s.

It is important to note that, in response to the widespread historiographic opinion of the Franco regime's animosity to foreign capital, the embassy expressly denied, also in the internal confidential reports, that the case could be interpreted as an attack on a foreign company. However, there were nuances in the British position. While the embassy, particularly in its trade department, was highly cautious, it denied discrimination on the grounds of origin and considered that no more could be done for the company. The central services of the Foreign Office deemed as intolerable that a British company was deliberately led to bankruptcy and recommended that the highest level of pressure be applied, although it consulted the Treasury. The Treasury agreed but considered that the ambassador should be correctly advised in view of the probable objections of the Spanish authorities to the rate of the company being 2.5 higher than that of the city council and one of the highest of Andalusia and that it was not acceptable to include the high costs of the London-based office. They could also argue that the early rescue would not have to involve a compensation to the company, given that it was operating without profits.

In order to prepare the negotiating strategy, representatives of the company, the Foreign Office and the Treasury met in London on 28 July 1952. At this stage, and aware of its weak position, the company only aspired to a modest economic compensation[5] and was willing for the compensation to be deducted from the net income derived from the price increase that it was requesting. Despite the British dealings with the Ministry of Foreign Affairs, the Ministry of Industry continued to refuse to authorise price increase until the company had reached an agreement with the city council, although it was evident that the latter was in no hurry to do so, leading to a dead end.

The Undersecretary of Industry confidentially informed the British commercial attaché that the reticence of the mayor to reach an agreement was due to local politics. Through its daily newspaper, the Phalange party in Seville launched a nationalist campaign against the company, alleging that the concession should be cancelled without compensation, as it had failed to fulfil the terms of the contract regarding the pressure and amount of water. Therefore, the company, as suggested by the manager, approved the hiring of the services of a Spaniard who was well connected with the local Phalange circles who, as well as receiving a commission would be authorised to pay bribes to facilitate the negotiations. These services proved to be unfruitful.[6]

Due to its accelerated financial decline and maybe as an element of pressure, the company communicated that in a few months it would not be able to pay the wages or continue providing the service. Within this situation of uncertainty, a representation of the employees of the company met in Madrid at the end of 1952 with ministerial representatives in order to transmit their concerns about the labour situation, which constituted an element of pressure on the Spanish authorities to speed up the resolution of the matter.

The embassy considered that Spain would not accept including the interests of the bonds or the expenses of London. The Treasury proposed alternatives of "creative accounting" such as inflating the authorised amortisation or the costs of Seville, including those of London. Despite the insistence of

5 The company was reticent to talk about specific figures, even with the Foreign Office. However, on a confidential basis, the manager informed the commercial attaché that in 1952 the Board of Directors had approved the claiming for 180,000 pounds (around 20 million pesetas). This figure was much lower than previously predicted.

6 This person could have been Antonio Gamero Martín, an administrative employee of the company, a councillor of the Unión Patriótica in Seville during the dictatorship of Primo de Rivera, chairman of the Colegio Pericial Mercantil and the Unión de Empleados de Escritorio, and who, surprisingly appeared in 1952 as a "representative" of the company undertaking different actions with the city council and the government.

the company, the Foreign Office gave precise instructions to the embassy to not enter into details regarding the economic compensation requested by the company. The city council needed the distribution network of the company but was less interested in the facilities in Clavinque as it had its own water resources. Therefore, the company placed its hope in them being acquired by the United States military authorities to supply the air base of Morón de la Frontera (Seville) within the context of the bilateral agreements of 1953. However, the United States considered it more appropriate to negotiate with the city council, not the company. Meanwhile, the Guadaira basin was running dry due to the increase in the extraction of water as a result of the expansion of irrigation, with an increasing number of wells and a higher extraction capacity of the pumping systems. Excluding evaporation and that absorbed by plants, the rainfall received was incapable of meeting the requirements, which was giving rise to a situation of water stress.

The Board of Directors resigned in 1952 due to its inability to reach sufficient price increases and open negotiations with the city council. Over the course of this year, there were several resignations and appointments of board members. The new Board of Directors favoured the adoption of a more belligerent position with respect to the city council, sending it an ultimatum. This clashed with the more prudent attitude of the manager and the recommendations of the embassy which, at this crisis point, received timely information from the consulate in Seville. The internal tension in the company intensified due to the accumulation of losses and the lack of perspectives for its future. The shareholders seemed to have assumed that they would not recover their capital as they had not received dividends since 1930. But now the novelty was that this idea was spreading to the bondholders who in the previous years were also subject to continual delays in the collection of their coupons and the amortisation of their securities. This led to their mobilisation and in the turbulent General Assembly of 1953, the elderly Scottish General Macpherson was elected. He possessed a large package of bonds but had no experience in foreign investment. He considered that this matter required firmness and efficiency and called for the reduction of the wages and pensions of the employees. The new president bombarded the manager, Adams, with bitter and continuous criticism and degraded him, which may have led to his resignation. This decision was taken by the Secretary Raine for the same reason.[7] The attitude of the new president aggravated the delicate situation of the company even further and

7 The board member Sparrow and A.G. Bibb, assistant to the manager, also resigned.

its internal division weakened its negotiating position with the city council. In the light of the above, the bondholder committee, created on 21 July 1953, obliged him to resign at the end of this year. Probably due to the influence of this committee, the moratorium on the payment of interests of the bonds, which had been extended since the 1930s and expired on 30 September 1953, was not renewed.

At the end of 1955, the negotiations, in which a representative of the bondholders also participated, accelerated, within a critical financial context for the company as the creditors (principally *Sevillana de electricidad*, the manufacture of the meters and the Ministry of Finance) increased their pressure. In February 1956, the company sent the city council a valuation of its assets undertaken by independent experts. The city council rejected it and proposed a price of 13,324,393 ptas (131,131 pounds), which was accepted by the company. The payments were to be made in pesetas, a third in cash and the rest in four equal annual payments with an interest of 4%. The city council took possession of the assets on 31 December 1956. It also acquired the warehouses of Seville and two lorries, which were not included in the concession and valued at 1.5 million pesetas, a third in cash and the rest in four years with interest. It took over the workforce in Seville in exchange for 500,000 ptas from the company. Also taking into account the net balance of the water sold by the company to the city council and of the filtered water sold by the city council to the company, the latter received a total of 10 million net pesetas from the Seville city council. The total received was only enough to pay the priority creditors and a small proportion of the bonds (6%) but was insufficient to amortise the bonds (5%) or the ordinary or preferential shares.

In 1957, the land of Alcalá de Guadaira was sold for 844,892 ptas. In the following year, an agreement was reached with the town council of Guadaira for the sale of the asset for 825,000 ptas: initial payment of 225,000 and three yearly payments of 200,000 ptas at 4% interest. While being much lower than the real value, the valuation was considered to be the best possible option. On 26 May 1959, the 73rd Ordinary General Shareholder's Meeting was held, which was presumably the last. In short, the company had to sell its assets for an amount that was much lower than their book value,[8] but this was the

8 The company would have received a total net amount of 11,669,892 ptas, equivalent to 106,090 pounds at the exchange rate used by the company in 1957 of 110 ptas/pound, although this amount was most probably lower due to the payments being spaced over four years and the deterioration of the exchange rate during that period. In 1956, the net book value of its fixed assets was 463,897 pounds.

least harmful alternative, as it could have been seized without compensation due to its inability to continue providing the service. The new municipalised services functioned poorly due to the lack of coordination between the technical and administrative sides, until the *Servicio Municipal de Aguas* was created in 1968 with its own budget and autonomous management (Aparicio 1990).

6 Conclusions

Seville Waterworks was the longest living foreign water company in Spain. It was a free standing company, a model that was frequently adopted by British companies operating abroad, which gave them great flexibility. Its financing (shares and debentures) came from the wide British capital market, the incipient "popular capitalism" which had begun to form at the end of the nineteenth century. Its shareholder, director and management structure were characterised by stability.

Throughout its long history, the company experienced several phases, which were common to all the public service companies. It had a golden era until the First World War. The continued, but assimilable, expansion of demand and the availability of financial and water resources enabled the service and distribution network to be expanded, reducing the unit costs due to the growing exploitation of economies of scale. Price stability allowed the rates charged to generate profits that could be distributed among the shareholders in the form of dividends. However, the deterioration of the exchange rate in certain situations, such as the Cuban war, negatively affected the results expressed in pounds.

The difficulty to obtain materials, the exchange imbalances and, particularly the inflation of costs due to the Great War and post-war period in an activity of a frozen and regulated unit revenue (rates) seriously affected the financial equilibrium of the company. In the 1920s, the company's response was to attempt to contain the costs by improving efficiency and modernising its facilities. The Depression and socio-political agitation of the 1930s led to an increase in defaults and wage demands. The political economy, foreign exchange and labour policy of the Franco regime constituted the final blow for the company. The high and prolonged inflation of costs (labour, energy, etc.) rendered the operations unfeasible, despite the continual rate increases which were delayed and insufficient. To this, we should add an exchange rate that was arbitrarily fixed by the government, and which had a highly negative effect on the results in pounds. As a backdrop, over the company's long lifespan, becoming worse in the final decades, was the exhaustion of the water resources of the company. In light of all of this, the only alternative was the municipalisation of the service, which would enable the joining of the greater water resources of the city council with the company's distribution network. The lack of financial resources and political will of the city council deliberately extended the negotiations in which the British embassy intervened. Finally, in 1957, the city council took possession of the facilities for a price that was significantly lower than their book value, seriously harming the bondholders and particularly the shareholders, who lost all of their capital.

7 Sources

Seville Waterworks, Annual Reports 1887–1956.

Public Record Office:

- *Seville Waterworks* 1891.
- *Seville Waterworks* 1945 FO 371-49585.
- *Seville Waterworks* 1952 FO 371-102025.
- *Seville Waterworks* 1953 FO 371-107694.

London Stock Exchange, Annual Reports 1884–1935.

Municipal Archive of Seville (Archivo Municipal de Sevilla):

- Aguas, box 1606, pieza 14. Extracto del expediente general de abastecimiento de aguas a esta ciudad, 1º-5º piezas.
- Aguas, box 1606, Informe sobre el abastecimiento de aguas a Sevilla, del ingeniero Luis Moliní, 1 June 1885.
- Aguas, box 1606, copia de la escritura general de abastecimiento de aguas, Sevilla, 1897.
- Aguas, box 1607, copia de la escritura de cesión de la concesión de Easton a *The Seville Waterworks*.
- Abastecimiento de aguas. Bases de la escritura de transacción, 18 February 1901, Tip. de La Andalucía Moderna, Sevilla, 1902.
- Aguas, expedientes y documentos, 1871–1901, no. 31, Contrato de ampliación del abastecimiento de agua, 26 October 1912.
- Colección legislativa, Aguas, box 874, Memoria sobre la municipalización del servicio de aguas potables, 1925.
- Colección legislativa, Aguas, box 874, Informe relativo al proyecto de abastecimiento de agua complementario de la ciudad.

Appendix

Board of Directors, 1882–1959

Director	Position/years	
Adam, John Mackay	M 1933–1959	D 1953–1959
Bithell, John J.	M 1908–1921	
Blaber, William	D 1891–1903	
Calzada, T. de la	D 1882–1885	
Castañeda,Tiburcio	D 1900–1910	
Compton-Bracebridge, J.E.	D 1882–1885	
Crew, Charles	S 1886–1894	
Curry, T.C.	M 1922–1924	
Devonshire, Easton	D 1896–1910	
Dickinson, Arthur	D 1909–1921	
Dickinson, William	D 1922–1935	P 1936–1947
Easton, James Rammel	D 1939–1951	
Emsheimer, Rudolf	D 1953–1956	
Evans, David	D 1882–1890	P 1891–1901
Filgate, Macartney E.	D 1917–1919	P 1920–1925
Friend Tynell, Charles A.	M 1882–1907	
Grace, Oliver Jelf	P 1953–1956	
Gray, George Wynter	D 1931–1941	
Hamilton, Gerard Montague	D 1933–1947	P 1948–1952
Hamilton, John Montague	S 1895–1925	D 1926–1932
Horner, Edward	P 1886–1890	
Jackson,Frank S.	D 1888–1921	
Johnston, E.F.	D 1882–1890	
Lambert, Cowley	D 1882–1901	P 1902–1917
Lawford, Evelyn Godfrey	D 1942–1952	
Mitchell, A.J.	S 1939–1945	
Mowat, Thomas Flett	S 1926–1935	
Raine, G.N.	S 1946–1959	
Schrimiton, George Ernest	D 1948–1952	
Segovia, G.	D 1882–1885	
Tufnell, Carleton F.	D 1920–1926	P 1927–1935
Walter, Vivian Lucas	D 1953–1959	
West, P.	M 1925–1932	

Board of Directors, 1882–1959 (*cont.*)

Director	Position/years
Wilson, T. Perceval	D 1902–1909
Young, Robert M.	S 1882–1885

P: President; D: Director; S: Secretary; M: Manager
SOURCE: *ANNUAL REPORTS* 1887–1956

References

Aparicio, D. e. a. 1990. *El agua en Sevilla*. Seville: Ediciones Guadalquivir.

Barrionuevo, A. 2005. *Sevilla: las formas de crecimiento y construcción de la ciudad*. Seville: Universidad de Sevilla, 2ª ed.

Braojos, A., M. Parias, and L. Álvarez. 1990. *Sevilla en el siglo* XX (Tomo 1 [1868–1950]). Seville: Universidad de Sevilla.

Broder, A. 1981. "Le rôle des interets économiques étrangers dans la croissance de l'Espagne au XIXe siècle, 1767–1924". These d'Etat, Universitè de Paris I.

Calvo, P. 1994. "El ocaso de la cultura del caño libre: protesta vecinal en Segovia en los albores del directorio civil (1925–1926)." *Investigaciones históricas: Época moderna y contemporánea* 14: 221–36.

Carreras, A., and X. Tafunell. (ed.). 2005. *Estadísticas históricas de España. Siglos XIX y XX*. Bilbao, Fundación BBVA.

Collantes de Terán, A. et al. (2011). *El agua y Sevilla. Abastecimiento y saneamiento*. Seville: Emasesa metropolitana.

Costa, M. T. 1981. "Iniciativas empresariales y capitales extranjeros en el sector servicios de la economía española durante la segunda mitad del siglo XIX." *Investigaciones Económicas* 14: 45–83.

Fernández Chaves, M. 2011. *Los caños de Carmona y el abastecimiento de agua en la Sevilla moderna*. Seville: Emasesa metropolitana.

González Cordón, A. 1984. *Vivienda y ciudad Sevilla 1849–1929*. Seville: Ayuntamiento de Sevilla.

González García, A. 1981. *El gas en Sevilla, 100 años de historia 1846–1945*. Seville: Artes gráficas salesianas.

Hassan, J. 1998. *A history of water in modern England and Wales*. Manchester: Manchester University Press.

Martínez-López, A. 2020. "Inversión extranjera y abastecimiento de agua: Algeciras, 1895–1952." *História Unisinos* 24 (2): 209–20. https://doi.org/10.4013/hist.2020.242.04.

Matés-Barco, J. M. 1997. "Las sociedades anónimas de abastecimiento de aguas en Andalucía: una primera aproximación." *Boletín del Instituto de Estudios Giennenses* 167: 103–130.

Matés-Barco, J. M. 2002. "Strategies of Foreign Firms in the Sector of Water Supply in Spain." In *Transnational companies, 19th-20th centuries*, edited by H. Bonin et al., 301–16. Paris: Plage.

Millward, R. 2007. "La distribution de l'eau dans les villes en Grande Bretagne au XIXe et XXe siècles: le gouvernement municipal et le dilemme des compagnies privées." *Histoire, economie, sociétè* 26 (2): 111–28. https://doi.org/10.3917/hes.072.0111.

Posadas, C. A. 1995. *Sevilla y el estado, 1892–1923: una perspectiva local de la formación del capitalismo en España* (Vol. 7). Seville: Universidad de Sevilla.

Stone, I. 1999. *The Global Export of Capital from Great Britain 1865–1914. A Statistical Survey.* London: Palgrave Macmillan. https://doi.org/10.1007/978-0-333-98377-5.

Wilkins, M. 1998. "The Free-Standing Company Revisited." In *The Free-Standing Company in the World Economy, 1830–1996*, edited by M. Wilkins, and H. G. Schröter, 3–64. Oxford: Oxford University Press.

CHAPTER 8

The New Supply of Potable Water in Malaga (Spain) in the 1920s: A Sustainable System?

Víctor Manuel Heredia-Flores

Abstract

The private management model in the water supply service in the city of Malaga ended with the municipalisation of 1913. The serious problems existing in the quality and quantity of the supply continued until the 1920s. Then the engineer José Bores proposed the complete renovation of the system under new bases: modern water supply without energy consumption, new urban distribution network, new sewerage network, solution for the industrial and agricultural use of Torremolinos springs, hydrogeological study of the aquifer, sanitary control of the water and establishment of an exploitation by means of meters. These aspects, especially the concern for the environmental preservation of the aquifer and its surroundings (creating a protection area by planting trees) and the comprehensive vision of the problem of supply and sanitation, give the Bores plan an undoubted modernity.

Keywords

Bores Romero – Malaga – municipalisation – sustainability – water supply

1 Water Supply and Sustainability

Public water providers face many challenges in meeting the needs and expectations of the communities they serve. Providing an adequate supply of safe water is still a preeminent goal of all urban water providers, but the nature of the challenges inherent to continuing to meet that goal has changed dramatically since the first public water supply systems were developed thousands of years ago. As public concerns about water continue to evolve in the twenty-first century, so do perspectives on what "sustainability" might mean for urban water supply systems.

 | DOI:10.1163/9789004679634_009

The creation of the first population centres forced the establishment of water supply systems from nearby rivers and springs. Long before the concept of sustainability was formulated, communities understood that they could only sustain themselves as long as their water uses stayed within the limits of local water supplies. The continuing growth in urban and agricultural demands has increasingly been met by importing water from distant sources. In recent decades, new technologies and infrastructure have further increased the water supply for cities, for example by recycling wastewater, desalination of ocean water, or harvesting rainwater. Strengthening water supplies and anticipating growing water demand has been, and will continue to be, a key aspect of the sustainability of supply systems (Richter et al. 2018).

The best way to use water resources sustainably is effective and comprehensive management. The increase in demand and the deterioration of natural conditions both quantitatively and qualitatively due to these increases in use and demand, and climate change negatively affect water resources and continue to put pressure on them. In the European Union, it is argued that water pricing policies implemented in combination with other non-tariff measures prove to be the most effective in reducing domestic water consumption. Regardless of water consumption targets, water pricing remains a key instrument to achieve cost recovery for water services to ensure maintenance and financing of existing and future water infrastructure (European Environment Agency 2016).

Water is a renewable natural resource of vital importance for the development of life and economic activities. Consequently, water management plays a key role in the design of sustainable development models. It is a resource that must be managed rationally and efficiently, with the aim of ensuring human health and the conservation of ecosystems.

The quantity and quality of water systems are determined by natural factors, such as geology and climate (particularly rainfall patterns), but also by pressure from human activities. The increase and overexploitation of catchments, the increase in demand, the contamination of aquifers, losses in pipes or wastewater treatment systems are just some of the challenges involved in a sustainable management model for water destined for to urban consumption (Guaita and Jiménez 2008).

The United Nations Development Programme (UNDP) establishes that water plays a fundamental role in sustainable development. Access to water and an effective sanitation system is a fundamental right of human communities. In addition, they are necessary to avoid poverty. But attention to the demand for water implies the protection of surface and groundwater and its environmental surroundings. Planning and management are tools to achieve these objectives, taking into account geological, climatic and environmental conditions.

Responsible use and the balance between consumption and recharging of supply sources are main objectives. Precisely one of the Sustainable Development Goals established by the UN is to guarantee access to sources of drinking water and sanitation for the entire world population.[1]

The main objective of a water supply system is to provide water in an adequate, safe and economically viable way, without health risks and without compromising access to the resource and its availability for future generations. The systems are made up of a series of technologies that cover the different functional phases, such as collection, treatment, storage, distribution and domestic use. Likewise, sanitation systems must ensure the safe management of wastewater from the point of generation to the point of discharge or reuse with their corresponding technologies.

Water supply and sanitation systems go hand in hand, and together they must protect and promote human health by providing a clean environment and by interrupting the cycle of disease. According to this, these systems must meet sustainability criteria such as: control over the risk of contamination at all points of a supply system; the impact on the environment, which implies the intensive consumption of energy and natural resources used in the construction, operation and maintenance of the system; adaptation to local circumstances through the use of technology and procedures designed according to these changing circumstances (population, territory, soil, climate); economic costs appropriate to the capacity of the communities; and the acceptance by the community of the need and common interest of these systems (Sustainable Sanitation Alliance 2008).

The historical perspective is revealed as a useful tool to understand the evolution of the forms of water management as a natural resource and as a public service in the last two centuries. In a region like Andalusia, where water is a scarce resource and subject to climatic conditions, the study of the past is a line of action to address current and future policies for environmental protection and sustainable management of natural resources. From the appearance and consolidation of the "Modern Drinking Water System" since the nineteenth century, a water supply model has been developed that is at the origin of the current systems of this urban service, characterised by a universal and permanent supply, subject to a tariff system and whose management is distributed between concessions to private companies and the different models of public management (Matés-Barco 1998).

---

1 About UN-Water: https://www.unwater.org/about-un-water/members-and-partners/united-nations-development-programme-undp.

2 Modernisation of Water Supply Systems

Water supply is a fundamental service intended to satisfy one of the primary needs of human activity. In Spain, its study has made notable progress in recent decades, largely thanks to the work of Professor Juan Manuel Matés-Barco through an extensive list of contributions in which he has been offering a very complete overview of the modernisation process of this basic service. But, as he himself has indicated, it is still necessary to go deeper into the study of local cases to establish more precisely the times and characteristics of this process (Matés-Barco 2014).

With industrialisation, water became a product of the market economy and, therefore, a commodity that required large investments and could bring large profits. Since then, the evolution towards a universal household drinking water supply system has presented notable changes from various points of view: technical, economic, political, legal and business. There has been a significant transformation in the way of conceiving the management of public services and in the relationships between the public and private spheres of action. The successive alternatives that have been adopted have affected both the applied technologies and the ways of managing an essential public service. Consequently, a line of work dedicated to the analysis of the role played by public and private initiatives in the modernisation of the supply service has been consolidated in economic historiography (Matés-Barco 2006, 45).

Matés-Barco has defined the classic drinking water system as the one existing in cities during pre-industrial times and at the beginning of industrialisation. Broadly speaking, this system was characterised by a scarce offer, a very limited organisation, the non-generalisation of home supply and the coexistence of collective forms of supply with other private ones. Another fundamental characteristic of the service was the total absence of sanitary controls and adequate sanitation systems, with which the sewage ended up contaminating the wells and underground springs.

This situation evolved due to the influence of several factors, such as the availability of new technology for hydraulic works, the pressure exerted by a continuously growing demand and the development of business forms of management more in line with the progressive expansion of capitalism. The final result, which will not be visible in most populations until well into the twentieth century, is the modern drinking water system, characterised by increased demand, privatisation of supply, home supply, the application of tariffs for consumption and the implementation of technology in the collection, transportation, distribution and sanitary treatment of water.[2]

The creation and modernisation of urban supply networks were linked, in a first stage, to the action of private capital through companies that hoped to do business with the new situation of the water market, subject to a notable increase in demand as a result of the processes of industrialisation, population growth and urbanisation. The concession system allowed public administrations to maintain ownership of the activity while at the same time transferring its management to private companies. In the European context, around the First World War, the trend towards greater administrative interventionism gained weight, which translated into the retreat of the concession system in favour of direct public management (Matés-Barco 2008, 195), in a way that that the municipalities, as responsible for the water supply, were assuming the management of the service and the execution of the necessary works, with the legal and financial support of the State (Matés-Barco 2009b, 24–7). Millward (2007) points out that, at least for the British case, the decrease in business opportunities and the drop in profitability explain this new situation.[3]

The construction and management of urban supply networks in the Spain of the Restoration, between 1875 and the 1920s, were closely linked to the participation of private capital due to the economic and technical weakness of local administrations (Simón and Matés-Barco 2010, 33–4). The provision of these services by the concessionaires suffered in many cases from a lack of supervision by the competent administrations, which allowed the private companies to do so and did not demand compliance with the clauses of the concession contracts, which generally enjoyed extensive terms that could reach almost a hundred years. This lack of concern, largely explained by the connivance of interests between dealers and local authorities, gave rise to permanent irregularities that affected consumers to the detriment. The local oligarchies saw the implementation of water supply projects as an excellent opportunity for their businesses (Matés-Barco 2008).

Sooner or later, this type of situation degenerated into conflicts between public and private interests that, during the first decades of the twentieth century, translated into a stricter control of the actions of the concessionaires (Simón and Matés-Barco 2010, 32–6). Progressively, the favourable trend

2 These concepts have been studied and developed by Matés-Barco (1998, 1999, 2008, 2009a, 2009b) and Simón and Matés-Barco (2010).

3 In the Italy of the period between the centuries there was a varied panorama when facing the modernisation of the supply systems, while in Portugal the different conditions of the service resulted in different evolutions in Lisbon and Porto (Bigatti 1997; Silva and Cardoso de Matos 2004).

towards the municipalisation of the management of public services gained weight, especially that of drinking water supply, one of the most sensitive due to its immediate repercussion on the quality of life of the population. The question of municipalisation was repeatedly raised in almost all the cities in which the service was controlled by private concessionaires, although its subsequent evolution followed different solutions and different rhythms depending on the respective municipal circumstances.

One of the first Spanish cities to lead the transition from the private management model to the public management model through municipalisation was Malaga. This city on the Andalusian Mediterranean coast had experienced a notable demographic growth in the middle decades of the nineteenth century, almost doubling its population in less than five decades: from 68,000 inhabitants in 1842 to 133,000 in 1887. This rise had coincided with a wide development of commercial and industrial activities, linked in the first case to the export of wines and fruits from the country and in the second to various manufacturing initiatives in the steel, chemical and textile sectors that turned the city into one of the most important centres of the first Spanish industrialisation (Parejo 2009).

At first, in the mid-nineteenth century, the local administration was unable to technically and financially assume the modernisation of the supply systems, with the consequent inability to adapt to the new and growing water needs posed by domestic consumers and industrial. In 1875 the Malaga City Council conceded the construction of infrastructures and the management of the service to a private company through a concession contract. This privatisation allowed progress in the modernisation of the supply and distribution network until the concessionaires lost interest in maintaining and improving the service, largely due to the diminishing returns in its operation. The proliferation of complaints about the poor state of the supply and a new political environment prone to the recovery of the municipal initiative in the provision of services ended up giving way to a third stage as a result of municipalisation in the second decade of the twentieth century, which it was a new impetus for the modernisation of the urban drinking water supply service.

3 The Water Supply to Malaga until 1920: Private Companies and Municipalisation

The water supply was a serious social problem in Malaga in the first three quarters of the nineteenth century. The poor maintenance of outdated and insufficient infrastructures aggravated the hygienic deficiencies of the supply system and facilitated the transmission through contaminated water of a large number of diseases, which could sometimes cause epidemics, such as cholera or typhoid fever (Castellanos 1995, 68–71). Throughout the nineteenth century, water supply ceased to be just a resource to satisfy minimum vital needs and came to be studied as an essential part of the health infrastructure and as an element that had repercussions on the quality of life of the inhabitants. This new approach continued to be accompanied by strong social contrasts, which were also expressed in the way of access to drinking water: while the majority of the population had to go to a public fountain, a privileged group had direct ownership of a certain volume of water that reached his own home through the two existing supply networks: a municipal one, known as Aguas de la Trinidad, which took water from nearby springs called Almendral del Rey and Arroyo de la Culebra and which dated from the sixteenth century; and another, the San Telmo Aqueduct, built on the initiative of Bishop Molina Lario at the end of the eighteenth century and which was supplied by the Guadalmedina River. In reality, we can speak of a mixed system with three sources of supply, since to the two mentioned we must add the still very large number of homes that were supplied from their own wells, in many cases with great possibilities of contamination due to their proximity to black pits.[4] This fact is especially evident in those areas of the city where there was no distribution network.

In 1852, the doctor Vicente Martínez carried out a study of the sources of supply in which he concluded that the allocation per person per day was slightly higher than two litres, for a population of about 80,000 inhabitants (Martínez 1852, 332–3).

There is no doubt that there is a close relationship between the growth of the urban population and the worsening of the problem of water supply destined to meet both the industrial demand (especially intense in the case of the textile and chemical sectors) as well as the domestic and public service demand (cleaning, watering gardens). The City Council has been looking for new alternatives since the fifties of the nineteenth century, which materialised in the option of the Torremolinos springs, about 15 kilometres to the southwest of the city. The existence of abundant natural water sources in the eastern part of the Sierra de Mijas has been known since ancient times, although their use was limited to supplying the small nearby villages such as Torremolinos or Churriana, the movement of several flour mills and the irrigation of the adjoining lands.

4 The serious hygiene problems caused by the existence of private wells contaminated by the proximity of cesspools have also been pointed out by Bordeaux in the mid-nineteenth century (Fernández and Pérez 2003, 337).

The project carried out in 1866 by José Morer Abril (who had participated in the design of the urban distribution of the waters of Canal de Isabel II in Madrid) and Joaquín Pérez de Rozas established for a population of 130,000 inhabitants and considering a daily provision of 100 litres per person, the amount of 13,130 cubic metres as the necessary flow to guarantee the supply of Malaga.[5] The project contemplated a double pipeline with a length of 15 kilometres, using cast iron pipe and a regulating tank with a capacity of 20,000 cubic metres.

When a part of the infrastructure had been built, the lack of funds made the commission in charge of the Torremolinos water supply request a report on the possible solutions from the engineer Sancha. To finish the work, he proposed to grant the exploitation of the aqueduct to a concessionaire, who should take charge of the pending debts and of the part of the work that still remained to be carried out. The City Council accepted a proposal made by Fréderic Gros, from Paris, to take charge of the city's water supply service, committing to assume the pending accounts with the contractors and to finish the works, in exchange for the 99-year concession of the water that had not been sold and the exploitation of the service, leaving 2,500 cubic metres at the disposal of the corporation so that it could attend to the neighbourhood sources and municipal services.

On 2 August 1875, the concession contract between the City Council and Gros was signed and the works resumed immediately.[6] After satisfactorily passing the driving test, the official inauguration of the pipeline took place on 19 June 1876 (Figure 8.1). That day marked a new stage in the history of water supply to the city, which for the first time was able to have an abundant and continuous supply. But, at the same time, a period was beginning in which the management of the service was riddled with irregularities and controversies and, in addition, it continued to drag many of the previous deficiencies: coexistence of distribution networks and sewage systems without being properly isolated, survival for large part of the population of the system of domestic wells and minimal investments in the maintenance of the infrastructure. It was, in any case, a first step towards a supply system that we can describe as modern and typical of a contemporary society.

5 The provision of 100 litres per inhabitant per day had served as a reference for Bordeaux very recently (Fernández and Pérez 2003, 339).

6 Archivo del Museo de Artes Populares de Málaga – Archive of the Museum of Popular Arts of Malaga – (hereafter AMAPM), box 71. *Contrato de concesión de construcción de obras del acueducto y distribución y aprovechamiento de Aguas de Torremolinos*. Malaga, 1875. Olmedo 1998, 52–63.

FIGURE 8.1 Official inauguration of *Torremolinos Waters* pipeline, 1876
SOURCE: MUNICIPAL ARCHIVE OF MALAGA

The formation of companies related to the supply of Malaga was always done abroad. This succession of transfers, bankruptcies, auctions, and companies took place in Paris and Malaga City Council hardly had any official knowledge of them. In any case, despite the name changes, the business remained in the same hands.[7]

Even so, the reports of the existing irregularities in the entire concession process, since 1875, were happening. In 1888 the supply concession was awarded at auction to Francisco Bergamín, a prestigious lawyer and politician who came to occupy various ministries in the last governments of the Spanish Restoration.

In the following years no investments were made in the maintenance of the basic infrastructures and only the minimum improvements were made that allowed the supply to continue or that could bring benefits to the entrepreneur, and the water from the springs was negotiated, playing with the characteristics of the aquifer. This is how the private galleries of San José, La Pellejera, and Rojas were excavated, which came to influence the decline of municipal springs. This covert privatisation of the waters of Torremolinos generated a strong controversy in the media that fuelled the demand for municipalisation by the progressive sectors of local politics (Heredia-Flores 2013 and 2018).

7 AMAPM, box 71. Olmedo 1989, 415–21.

The irregularities and breaches of the concessionaires, such as the non-construction of the regulating tank foreseen in the initial project, gave rise to frequent claims by users and public opinion. The issue of water supply was debated many times in municipal meetings, dealing on several occasions with proposals to recover the concession.

The municipalisation of the water service is explained by the new conditions of municipal politics in Malaga from 1909, when a majority of Republicans and Socialists formed (Arcas-Cubero 1985, 420–1). In 1913, the City Council approved the rescission of the concession contract,[8] with broad support in local society. Both parties, the City Council and the company, agreed to submit to the opinion of a commission made up of three lawyers of national prestige: Gumersindo de Azcárate, Antonio Maura and Luis Díaz Cobeña. These jurists issued their opinion in 1915, which granted the City Council the powers for the free administration and disposal of the waters of Torremolinos.[9]

The end of the private concession, not so much because of the shortcomings of the service as because of the strong political rivalries within the local bourgeoisie, presents a suggestive parallelism with a contemporary but distant model. The municipalisation of the supply to the city of Los Angeles, in California, in 1902 was the result, according to Mackillop (2004), of a commitment by local elites who, in an apparent paradox, opted for public management because it served their interests better than the previous private company.

The hopes raised by the municipalisation of the service were not justified in the years immediately following. The projects to install a new pipeline and build a regulating tank, which would allow water to be distributed in the neighbourhoods and higher floors without the need to resort to motors or pumps, had to wait while the legal situation of the service was resolved and, also, to because of the lack of resources of the municipal treasury. In fact, the trajectory of the water supply in the capital of Malaga was related to the general situation of the service in the large Spanish cities. In most of them, concessions had been made to private companies that were showing obvious signs of inefficiency: fraud, non-compliance with contracts, little or no renewal and improvement of facilities and equipment.

At the beginning of the twentieth century, a series of professionals (engineers, architects, lawyers…) were incorporated into municipal politics, who contributed new ways of managing town councils and a greater concern for solving the serious problems they suffered in their work as councillors and

8 AMAPM, box 71, 33–37. *La Unión Mercantil*, 11 July 1913. Olmedo 1989, 420–30.

9 Archivo Municipal de Málaga – Municipal Archive of Malaga – (hereafter AMM). *Boletín Municipal de Málaga*, 6 July 1915.

technicians. the cities. In this way, the theory and practice of municipalisation spread. The municipalisation was justified by the need to universalise the supply, carry out the necessary works for this purpose and improve the provision of the service (Matés-Barco 1998, 195).

4 The New Aqueduct of Torremolinos Waters: Bores Romero's Project

At the beginning of the 1920s, the city of Malaga presented an urban infrastructure characterised by the great limitations of the water supply and sanitation networks. The demographic growth of the first decades of the century had been aggravating several problems that, in addition, there were long periods of neglect and abandonment. Between 1920 and 1930 the population increased by 30,000 people, reaching 180,000 inhabitants. The water supply was insufficient, and the sewage network accumulated deficiencies and constituted a huge cesspool, into which sewage, garbage and all kinds of waste were dumped. For these reasons, the City Council decided to create the Roads and Works Headquarters, which was occupied, after a competition, by the engineer José Bores Romero[10] in March 1921. The supply service came to technically depend on this Headquarters. Bores was immediately entrusted with the creation of a sanitation plan for the city, which he carried out in a short time and which he presented with the title of "General Organic Plan for the Complete Sanitation of the City".

The water supply system was still deficient. The nearly 24,000 cubic metres per day available to the City Council were insufficient for the existing population (the theoretical allocation per inhabitant was 160 litres per day, without taking into account system losses). The cast-iron pipe was nearly half a century old, and breaks caused constant service interruptions. According to Bores, of the aforementioned amount of available water, about 13,000 cubic metres were not used, that is, more than half, which was lost in the intake itself or due to leaks in the pipe. On the other hand, for 10,427 buildings that existed in the urban area, only 4,816 outlets were in service, which meant that almost 54% of the properties did not have access to distribution and their neighbours had to go to public sources or wells in the courtyards of the houses (Bores-Romero 1921–1922).

10 Bores (Antequera, 1860-Zaragoza, 1936) was the nephew of the conservative politician Francisco Romero Robledo, who died in 1906. Throughout his long professional career, Bores covered practically all the fields of engineering of his time: the hydroelectric industry, railways, roads, water supply, sanitation, bridges, ports... (Heredia-Flores 2015).

Bores presented in December 1921 his project for "Improvements and reforms in the drinking water supply of Malaga", and three months later the project of "General Sewerage of the city", to complete the sanitation network. With regard to supply, the engineer estimated that, for a population of between 150 and 200,000 inhabitants, the city should have 40 to 50,000 cubic metres of water per day, which guaranteed an average of about 250 litres per person per day. After reviewing the shortcomings of the service, he concluded that it was necessary to "dispense with everything that existed and had been created and enter resolutely into the new era that the current and always successfully applied technique imposes". The project consisted of three phases: expropriation of private springs and necessary adjacent land, construction of a new conduit and formation of an adequate distribution network. The budget approved by the City Council in January 1923 for this project was somewhat less than eight million pesetas (Bores-Romero 1921–1922). The City Council announced a tender to jointly contract the execution of the works and the placement of the bonds to be issued. Of the two offers that competed, the one presented by *Banca Marsans S.A.*, of Barcelona, and *Empresa General de Construcciones*, of Madrid, was chosen. On 20 September 1923, the award contract was signed (Copia 1923).

Before the works began, Bores Romero requested that the Geological Institute of Spain to prepare a geological report on the hydrological basin of the Sierra de Mijas and, especially, on the area of the springs. The engineer Enrique Dupuy de Lôme, after carefully studying the Sierra de Mijas, came to the conclusion that all the aquifers of that limestone mountain massif were connected, so that any new spring would have repercussions in the decrease of the existing springs. In addition, Dupuy proposed a set of measures to protect the maintenance of the flow of the Torremolinos springs, advising that a protection zone be established (Dupuy 1923). This opinion allowed Bores to justify the expropriation of the private springs of San José and La Pellejera, of the easements of some irrigations and of the seventeen mills that still used the canal that ran through the village of Torremolinos to the beach. A final measure of protection was the expropriation of some 600 hectares above the springs and of a strip of land 14 metres wide for the layout of the canal.[11] These measures made it possible to release a flow of 23,000 cubic metres of Torremolinos water from other uses, with the aim of obtaining the largest amount possible for supplying the city. While the works were being carried out, the municipality of Torremolinos was integrated into that of Malaga, in 1924.

11 AMAPM, box 71, 40–1 and 44.

The works were carried out for almost three years, until in 1926 the new pipeline arrived in the city. The new aqueduct, whose route was different from that of 1876, comprised a total length of 18,430 metres, of which 2,554 metres corresponded to tunnels. The difficulties presented by the layout were overcome by a series of fifteen reinforced concrete bridges and aqueducts, highlighting the large metal bridge that crossed the Guadalhorce River, 216 metres long (Figure 8.2). Thanks to the design of the layout, the water moved by gravity, and it was not necessary to resort to motors. The sums invested in expropriations and works amounted to about eight million pesetas; an amount very close to the initial forecasts (Bores-Romero 1926).

Two regulating reservoirs were built near the urban area: in La Florida and in Olletas, with a capacity of 18,500 and 45,000 cubic metres, respectively (Figure 8.3). The first was completed at the beginning of 1927. The second, still under construction in 1929, would be responsible for supplying the upper part of the city.

With the execution of this project, the engineer Bores believed that the proposed objectives had been achieved: to obtain full and absolute control of a flow of 32,000 cubic metres per day, to effectively protect the springs, to resolve the issue of easements for the irrigators, create a canal with sufficient capacity to transport a larger volume of water in the future, ensure pressure so that the water reaches the highest houses and, finally, put the supply in conditions to establish a remunerative exploitation, for which it was the implementation of

FIGURE 8.2 Bridge over the Guadalhorce River
SOURCE: ARCHIVE OF THE MUSEUM OF POPULAR ARTS OF MALAGA

FIGURE 8.3 Olletas water reservoir under construction
SOURCE: ARCHIVE OF THE MUSEUM OF POPULAR ARTS OF MALAGA

the subscription by metre was essential (Bores-Romero 1926). After Bores left his position in the City Council, an extraordinary new budget of almost three million pesetas was approved to finish the works. However, the laying of pipes throughout the city to form the new distribution network was not carried out (Bello, Cabellos, and Núñez 1927).

Torremolinos new water supply project was intended to reduce supply deficiencies in a city that was growing both in number of inhabitants and in size, with the aim of making water reach the maximum number of consumers in their own homes. In addition, at that time the concern for the sanitary situation of the water network was already recognised as an essential part of the quality of the supply (Castellanos 1995, 73) and the Municipal Laboratory began to carry out regular bacteriological analysis of the water supplied (Bores-Romero 1926, 30).

The realisation of the great project of the engineer Bores Romero did not serve to solve the problem of water supply, although it allowed the service to be maintained at minimum levels. The executed works of the Bores plan were incomplete, since one of the most important parts remained to be carried out: the expansion and renovation of the urban distribution network, which was half a century old and suffered continuous breakages with the following loss of flow in the network and with a high risk of contamination. In addition, the flow supplied was still insufficient for the needs of the population, so the problem of water supply continued to be one of the most important issues in the city in the following decades.

5 Municipalisation and Modernisation: A Balance

The first phase of the modernisation of the water supply system to the Malaga city was possible thanks to the contribution of private capital of foreign origin (French in this case), given the financial and technical incapacity of the City Council. The Parisian businessman Gros went into business that was sensed with the speculation of the amount of water whose rights could be sold to individuals. This intervention of foreign capital managed, in effect, to finish the works and start up a distribution network, but it prevented the completion of some details of the project that weighed down the operation of the system during the following decades, such as the construction of regulating tanks.

Once the salable parts of the granted water were disposed of, the financial problems of the successive concessionaire companies began. The events that took place in the 1880s revealed the manipulation of the service by sometimes conflicting interests, without the City Council having the capacity or the will to intervene. The scandals derived from these speculative practices and the increasingly serious deficiencies of the service due to the lack of investment and the reduction of flows brought the "water business" to the centre of the political debate in the City Council, controlled between 1909 and 1915 by the alliance between Republicans and Socialists. The end of the private management model with the municipalisation of 1913 did not have immediate effects on the definitive modernisation of the service, which would still take a few years to arrive. The municipal decision to hire the veteran and prestigious engineer José Bores Romero was key, who immediately proposed an ambitious plan of action that included several points: drafting the projects for a new supply and a new sewer system (assuming the importance of the interrelation between both systems), resolution of the residual uses (industrial and agricultural) that still weighed on the springs and hydrogeological study of the aquifer to establish the influence of the adjoining sources and create a protection zone. These aspects, especially the concern for the environmental preservation of the aquifer and its surroundings and the global vision of the problem of supply, from its sources to distribution, give Bores' plan an undoubted modernity.

However, the inauguration of the new supply in 1926 did not alleviate the social differences in access to water, nor was it completed with the essential renewal of the distribution network, shortcomings that continued to hamper the service despite the notable technical and financial effort deployed by the

City Council. It can be said that the change in the management model implied an authentic but partial modernisation of the system and the overcoming of speculative activities regarding supply, which continued to be insufficient to satisfy the social and urban demands of a growing city.

Bores Romero's project showed a clear interest in the environmental aspects of supply, establishing a comprehensive treatment for the protection of the environment of the sources, the construction of an aqueduct according to the most modern technology of the time (but avoiding the use of motors, since the water moved by gravity, which meant energy savings), the provision of pressure regulating tanks and the design of a complete distribution network and a sanitation network. It was not completed, and the lacks were not fully resolved, but it was a clear precedent for planning a sustainable supply system.

6 Archival Material

Archivo del Museo de Artes Populares de Málaga (Archive of the Museum of Popular Arts of Malaga – AMAPM –):
- Box 71.
- *La Unión Mercantil*, 11 July 1913.

Archivo Municipal de Málaga (Municipal Archive of Malaga): *Boletín Municipal de Málaga*, 6 July 1915.

References

Arcas-Cubero, F. 1985. *El republicanismo malagueño durante la Restauración (1875–1923)*. Cordoba: Ayuntamiento de Córdoba.

Bello, S., J. Cabellos, and J. Núñez. 1927. *Dictamen sobre la recepción de las obras de la nueva traída de aguas de Torremolinos*. Malaga: Ayuntamiento de Málaga.

Bigatti, G. 1997. "La conquista dell'acqua. Urbanizzazione e aprovvigionamento idrico." In *L'acqua e il Gas in Italia. La storia dei servizi a rete, delle aziende pubbliche e della Federgasacqua*, edited by G. Bigatti et al., 25–161. Milano: Franco Angeli.

Bores-Romero, J. 1921–1922. *Proyectos de mejoras y reformas en el abastecimiento de aguas potables y de nuevo alcantarillado general de la ciudad*. Malaga: Ayuntamiento de Málaga.

Bores-Romero, J. 1926. *Nueva traída de aguas de Torremolinos. Memoria sobre las obras*. Malaga: Ayuntamiento de Málaga.

Castellanos, J. 1995. "Enfermedad e infraestructura sanitaria: Una aproximación histórica al abastecimiento de aguas de la ciudad de Málaga." *Péndulo* 7: 68–73.

Copia. 1923. *Copia de la escritura otorgada por el Excmo. Ayuntamiento de Málaga con la Banca Marsans S.A. de Barcelona y la Empresa General de Construcciones de Madrid, para realizar el proyecto de mejora en el abastecimiento de aguas potables*. Malaga.

Dupuy de Lôme, E. 1923. *Estudio del Instituto Geológico de España acerca de la cuenca hidrológica de la Sierra de Mijas o Torremolinos*. Malaga: Ayuntamiento de Málaga.

European Environment Agency. 2016. European water policies and human health. EEA Report no 32. Accessed 11 September 2022. https://www.eea.europa.eu/publications/public-health-and-environmental-protection.

Fernández, A., and P. M. Pérez Castroviejo. 2003. "El agua y la ciudad: Burdeos y Bilbao (1850–1920). Normas técnicas, normas sociales." *Zainak. Cuadernos de Antropología-Etnografía* 23: 335–50.

Guaita-García, N., and L. M. Jiménez-Herrero. 2008. "Agua y sostenibilidad." *Indice: Revista de Estadística y Sociedad* 28: 14–7.

Heredia-Flores, V. M. 2013. "Municipalización y modernización del servicio de abastecimiento de agua en España: El caso de Málaga (1860–1930)." *Agua y Territorio / Water and Landscape* 1: 103–118. https://doi.org/10.17561/at.v1i1.1038.

Heredia-Flores, V. M. 2015. "José Bores Romero, un ingeniero en la España de la Restauración." In *Pensar con la historia desde el siglo XXI: actas del XII Congreso de la Asociación de Historia Contemporánea*, edited by P. Folgueras et al., 1781–99. Madrid: UAM Ediciones.

Heredia-Flores, V. M. 2018. "El abastecimiento de agua en Málaga (1860–1930). De negocio privado a servicio público." In *Agua y servicios públicos en España y México*, edited by J. M. Matés-Barco, and J. J. P. Rojas-Ramírez, 121–53. Jaen: UJA Editorial.

Mackillop, F. 2004. "The Influence of the Los Angeles "Oligarchy" on the Governance of the Municipal Water Department, 1902–1930: A Business Like Any Other or a Public Service?" *Business and Economic History On-Line* 2. Accessed 26 March 2022. https://thebhc.org/index.php/influence-los-angeles-oligarchy-governance-municipal-water-department-1902-1930-business-any-1.

Martínez y Montes, V. 1852. *Topografía médica de la ciudad de Málaga*. Malaga.

Matés-Barco, J. M. 1998. *Cambio institucional y servicios municipales. Una historia del servicio público de abastecimiento de agua*. Granada: Comares.

Matés-Barco, J. M. 1999. *La conquista del agua. Historia económica del abastecimiento urbano*. Jaen: Universidad de Jaén.

Matés-Barco, J. M. 2006. "Las empresas de abastecimiento de agua en Andalucía (1840–2000)." In *El debate del agua en Jalisco y Andalucía*, edited by J. R. Navarro and J. Regalado, 45–93. Sevilla: Junta de Andalucía.

Matés-Barco, J. M. 2008. "Empresas, sociedad y servicios públicos: del Estado prestador al Estado regulador." *Revista Empresa y Humanismo* 11–1: 187–230. https://doi.org/10.15581/015.11.33297.

Matés-Barco, J. M. 2009a. "Las sociedades anónimas de abastecimiento de agua potable en España (1840–1960)." *Revista de Historia de la Economía y de la Empresa* 3: 177–218.

Matés-Barco, J. M. 2009b. "El desarrollo de las redes de agua potable: Modernización y cambio en el abastecimiento urbano." *Agenda Social* 3–1: 23–51.

Matés-Barco, J. M. 2014. "Las empresas concesionarias de servicios de abastecimientos de aguas potables en España (1840–1940)." *TST. Transportes, Servicios y Telecomunicaciones* 26: 36–67.

Millward, R. 2007. "La distribution de l'eau dans les villes a Grand-Bretagne au XIXe et XXe siecles. Le gouvernement municipal et le dilemme des compagnies privées." *Histoire, Economie, Societé* 26 (2): 111–128. https://doi.org/10.3917/hes.072.0111.

Olmedo Checa, M. 1989. *Miscelánea de documentos históricos urbanísticos malacitanos.* Malaga: Ayuntamiento de Málaga.

Olmedo Checa, M. 1998. *José María de Sancha.* Malaga: Benedito Editores.

Parejo Barranco, A. 2009. *Historia económica de la provincia de Málaga (1833–2008).* Malaga: Diputación Provincial de Málaga.

Richter, B. D., M. E. Blount, C. Bottorff, H. E. Brooks, A. Demmerle, B. L. Gardner, H. Herrmann et al. 2018. "Assessing the sustainability of urban water supply systems." *Journal AWWA* 110 (2): 40–47. https://doi.org/10.1002/awwa.1002.

Silva, A. F., and A. Cardoso de Matos. 2004. "The Networked City: Managing Power and Water Utilities in Portugal, 1850s–1920s." *Business and Economic History - On Line* 2: 1–45. Accessed 12 August 2022. https://thebhc.org/networked-city-power-and-water-utilities-portugal-1850s-1920s-0.

Simón, I., and J. M. Matés-Barco. 2010. "El abastecimiento de agua potable en México y España: Cambio institucional y aparición de las empresas privadas (1870–1930)." In *Agua, territorio y medio ambiente. Políticas públicas y participación ciudadana,* edited by J. R. Navarro, J. Regalado, and A. Tortolero, 21–49. Guadalajara: Universidad de Guadalajara and Seminario Permanente ATMA.

Sustainable Sanitation Alliance. 2008. "Towards more sustainable sanitation solutions". Accessed 4 September 2022. https://www.susana.org/en/knowledge-hub/resources-and-publications/library/details/267.

CHAPTER 9

Water Supply in the El Aljarafe District (Seville Province, Spain): A Historical Perspective

Jesús Raúl Navarro-García

Abstract

Despite the number of studies on water supply and treatment in rural Spain to date being quite small, this study addresses the situation in a district that is very close to Seville, the capital city of Andalusia. Although the district under study practically forms part of the city's metropolitan area today, it was almost completely rural up to the middle of the last century. Previously, the water supply had been characterised by the use of own resources, but demographic growth and greater pressure from human activities revealed the evident failings of a system based on the use of springs and public and private wells. The search got underway for outside water resources and joint management was instigated with the backing of Seville Provincial Council, ultimately resulting in the setting up of a joint municipal company called ALJARAFESA. ALJARAFESA manages the water for 30 small and medium-sized localities with populations that range from barely six hundred inhabitants to over forty thousand in a widespread area with major pumping requirements and many kilometres of pipes.

Keywords

Aljarafe – ALJARAFESA – Seville – water resources – water supply

1 The El Aljarafe Occupation Process: The Increase in Demographic Density and Its Impact on Water Supply and Sanitation from the Mid-Twentieth Century

El Aljarafe is a plateau delimited by the Guadalquivir River to the east, the Guadiamar River to the west, the Guadalquivir marshes to the south and, to the north, the Campo de Gerena district and the foothills of the province's Sierra Norte mountains (Tahiri 2011, 115; García 2016).

 | DOI:10.1163/9789004679634_010

The water is currently supplied to the populations in the district by a joint municipal water company created in 1971 called ALJARAFESA. This municipal service company is run by 19 local councils in the district including the Seville Provincial Government, which also endorsed its creation and had initiated public water supply in the district a year previously, in 1970. ALJARAFESA provides water supply and sanitation services to 30 towns in the province and a total of approximately 320,000 inhabitants (2021). The majority of these towns are in the El Aljarafe district, along with a few other municipalities such as Gerena, in the Campo de Gerena district; La Algaba, in the La Vega district, and more recently, Isla Mayor, in the Guadalquivir marshes. These last three municipalities (Gerena, La Algaba and Isla Mayor) mark the current boundaries of the area supplied by ALJARAFESA.[1] The provincial border with the province of Huelva marks the western boundary while the municipalities of Aznalcóllar, Gerena, La Algaba, Tomares, Gelves, Palomares del Río, Isla Mayor, Villamanrique de la Condesa, Pilas, Carrión de los Céspedes and Castilleja del Campo complete the circle. The most distant points from the provincial capital are marked by the localities of Aznalcóllar to the north, Isla Mayor and Villamanrique de la Condesa to the south and Castilleja del Campo to the west, which are all approximately forty km away.

The entire area stands out for its historical dependence on the city of Seville, not only during the Muslim period, when it, and another twelve rural districts, formed the *cora sevillana* (a kūra, a territory of the Caliphate of Cordoba), but also much later, when authority over El Aljarafe came under royal administration, the Sevillian Church, and, at various other times, under some military, noble, or manorial orders (Tahiri 2011, 115; Rodríguez 2019, 136; Herrera 1981; 2010, 21–33). Proximity was not the only reason for these historical ties: the district's historical dependence on the city of Seville was also economic, social, judicial, and religious (Borrero 2005, 60).

Population distribution within the province of Seville is marked by a strong demographic concentration around the city of Seville itself, with 70% of the population occupying 10% of the surface area. Given its short distance from Seville, clustering can also be found in the area supplied by ALJARAFESA. ALJARAFESA serves a surface area of approximately 1,647 km^2 and includes some very extensive municipalities such as Aznalcázar (the sixth largest in Seville province at 449.84 km^2), as well as other smaller towns and villages such as Aznalcóllar (198.96 km^2), Sanlúcar la Mayor (135.41 km^2), Gerena (129.9 km^2) and Isla Mayor (114.38 km^2). These five municipalities are amongst the furthest

1 Aznalcóllar, Isla Mayor and Villamanrique are the last localities to have come under the joint management of ALJARAFESA.

from the provincial capital and together represent 62.4% of the area supplied by ALJARAFESA, which indicates that the further from the city of Seville a municipality is, the lower its demographic density. The remaining municipalities supplied by ALJARAFESA are much more modest in size and, on average, qualify as small towns and villages: Villanueva del Ariscal (62.48 km^2), Bollullos de la Mitación (62.36 km^2), Villamanrique de la Condesa (57.67 km^2), Huévar del Aljarafe (57.58 km^2) and Salteras (57.46 km^2). All the localities smaller than Pilas (45.94 km^2) and Olivares (45.53 km^2) are in the bottom provincial quartile and occupy areas under 33 km^2 (Benacazón, 32.16 km^2; Valencina de la Concepción, 25.14 km^2 and Espartinas, 22.74 km^2). Half of the municipalities that ALJARAFESA supplies with water have very small surface areas below 20 km^2 (eight localities) or even 10 km^2 (another seven). These are all near the capital and have high demographic densities. All of this clearly has implications for such things as water supply and sanitation costs. Lastly, fifteen of the nineteen least extensive municipalities in Seville province are supplied by ALJARAFESA, with very small areas such as Castilleja de Guzmán (2.06 km^2), Castilleja de la Cuesta (2.23 km^2), Gines (2.9 km^2), Tomares (5.17 km^2), Carrión de los Céspedes (6.01 km^2), and Gelves (8.18 km^2).

Scattered settlements have existed in the El Aljarafe area since Roman (Ordóñez and Beltrán 2005) and Muslim times, with the irresistible expansion of human occupation driven by the availability of water. This is proven by the erstwhile existence of small towns and villages such as Castilleja de Talhara and Gelo (Hernández 2005), which disappeared with time, and others that would later give rise to present-day settlements such as Palomares[2] and Valencina de la Concepción. The latter developed out of two farms which, along with nearby farmsteads, were examples of the large estates that prevailed in the seventeenth and eighteenth centuries and coincided with manors and Crown properties in the district (Valiente and Martínez 2005; Iglesias 2005, 100; Zurita 2017).

Occupation of the El Aljarafe area continued to expand during the tenth century with over two hundred Moorish farmsteads, some of which might have included as many as two hundred dwellings (Tahiri 2011; El Faiz 1991). Although the population in the study area contracted during the fourteenth century, the following century saw great growth (González 1982). This resulted in the area having over thirty thousand inhabitants and a settlement network very similar to that of today, with the population regrouping in localities such as Sanlúcar la Mayor (with 500 inhabitants), Aznalcázar (with 400) and many

2 On which depended the future municipalities of Mairena del Aljarafe and Almensilla (Martín 2010, 35–7).

other nuclei with between one hundred and three hundred inhabitants. In the mid-eighteenth century, the population had once again shrunk and barely surpassed twenty thousand, although a sharp demographic rise began in localities such as Carrión de los Céspedes, Villamanrique, Pilas and Benacazón. However, this increase was not observed in other places such as Gelves and Palomares, and some localities such as Huévar and Aznalcázar even experienced demographic declines that were quite sharp in some cases. In any case, by the mid- nineteenth century, demographic densities did not exceed 15 inhab./km^2.[3]

Nowadays, the situation bears little resemblance to what has been described above. Urbanisation and demographic growth increased rapidly from the second half of the twentieth century and triggered "rapid metropolisation" (Cornejo 2007, 48), turning El Aljarafe into a key part of the Seville urban agglomeration. The territorial dispersion of the 30 urban nuclei currently supplied by ALJARAFESA means that distribution costs are high, although this is compensated for in part by the high demographic density along the north-south urban axis, where there is a compact cluster of municipalities (García 2014), all of which are under ten km from the capital: Castilleja de la Cuesta (8,300 inhab./km^2), Tomares (4,834 inhab./km^2), Gines (4,607 inhab./km^2), Mairena del Aljarafe (2,568 inhab./km^2), Bormujos (1,776 inhab./km^2), Castilleja de Guzmán (1,437 inhab./km^2), and Gelves (1,230 inhab./km^2). These municipalities (combined population 138,705) are included in the inner metropolitan ring, an urban continuum that includes some other localities with densities of over one thousand inhabitants/km^2, such as Santiponce (1,061 inhab./km^2), which is several kilometres from the aforementioned axis, and also Palomares, which has a relatively high demographic density of 647 inhab./km^2. It must be borne in mind that, in 2007, population density in the province was 132 inhab./km^2, so these urban nuclei are well above the provincial average¬ (which already exceeds the national average of 90 inhab./km^2). Of the ten most densely populated provincial towns, seven are supplied by ALJARAFESA: as mentioned above, Castilleja de la Cuesta, Tomares, Gines, Mairena del Aljarafe, Bormujos, Castilleja de Guzmán, and Villanueva del Ariscal. The last of these is located in the middle metropolitan ring, where densities fall considerably given that only Villanueva del Ariscal exceeds one thousand inhab./km^2 (1,400), despite its population of 6,673 inhabitants. The density in the other towns and villages on this ring is less than half of this: Umbrete (730 inhab./km^2), Espartinas (679 inhab./km^2), and Almensilla (414 inhab./km^2). The

3 These data refer not only to the localities mentioned but also to other localities in the wetlands that are not part of El Aljarafe proper (Iglesias 2005, 88–90) and have been included to calculate the density.

densities of all the other localities are even more modest: Albaida del Aljarafe (285 inhab./km^2), Benacazón (223 inhab./km^2) and Bollullos de la Mitación (168 inhab./km^2). Sanlúcar la Mayor (100 inhab./km^2), Olivares (92 inhab./km^2), and Salteras (35 inhab./km^2) have densities more in line with localities located on the outer metropolitan ring, the furthest from the provincial capital. On the outer ring, Carrión de los Céspedes barely surpasses a hundred inhab./km^2 (105 inhab./km^2) and the figures for Villamanrique de la Condesa (77 inhab./km^2), Huévar (50 inhab./km^2), Castilleja del Campo (39 inhab./km^2), and Aznalcázar (10 inhab./km^2) are extremely low. The densities for Gerena, Aznalcóllar, and Isla Mayor, which are not part of the El Aljarafe district but are also supplied by ALJARAFESA, are similar to those of the localities in the outer ring: 57 inhab./km^2, 30.7 inhab./km^2, and 51 inhab./km^2, respectively. Despite the low densities of some of these localities, only Aznalcázar ranks amongst those with the lowest densities in the province. In contrast, the neighbouring locality of Pilas, with 304 inhab./km^2, has a density comparable to those in the middle metropolitan ring.

Demographic growth of the entire population of Seville province was spectacular during the twentieth century: a cumulative 1.14%/year, which far exceeded that of Andalusia (0.73%) and Spain as a whole (0.77%). While affecting the provincial capital especially, this growth also involved the metropolitan area (Dos Hermanas, Alcalá de Guadaira) and many of the localities in the El Aljarafe district, such as Mairena del Aljarafe with 46,895 inhabitants, Tomares (25,370), Bormujos (22,390), and Espartinas (16,048), thus sparking an overspill effect. The increase of 1,175,000 inhabitants in the provincial population was practically generalised as it impacted all the municipalities except those in the Sierra Norte mountains and the southeastern part of the province. This is in stark contrast to the situation at the beginning of the twentieth century when six towns and villages in the El Aljarafe area that are currently supplied by ALJARAFESA were amongst the ten least populated in the province: Isla Mayor (152 inhabitants), Castilleja de Guzmán (171), Palomares del Río (433), Castilleja del Campo (461), Albaida del Aljarafe (634), and Tomares (690). The situation has changed greatly as, today, only one of the places supplied by ALJARAFESA is among the least populated in the province: Castilleja del Campo. Similarly, in 1900, none of the localities currently supplied by ALJARAFESA were among the ten most populated, whereas Mairena del Aljarafe is now home to the sixth-largest population in the province. Indeed, in 1900, the ranking of towns and villages did not resemble that of today: only three of the towns now supplied by ALJARAFESA exceeded (and only by a very small margin) four thousand inhabitants (Sanlúcar la Mayor, Pilas and Aznalcóllar); another three exceeded three thousand: La Algaba, Olivares, and Villamanrique, and another six, two thousand: Carrión de los Céspedes, Benacazón, Gerena, Villanueva del Ariscal, Umbrete, and Bollullos de la Mitación. Therefore, only twelve localities in all surpassed two thousand inhabitants and none of these were part of the inner metropolitan ring. However, the population density was already relevant in towns such as Castilleja de la Cuesta (819 inhab./km^2), Villanueva del Ariscal (545 inhab./km^2), Carrión de los Céspedes (497 inhab./km^2), Gines (387 inhab./km^2), La Algaba (221 inhab./km^2), Santiponce (198 inhab./km^2), and Umbrete (188 inhab./km^2), to a great extent due to their small metropolitan surface areas, above all in the cases of Castilleja de la Cuesta, Gines, Carrión de los Céspedes, Santiponce, and even Umbrete. At the beginning of the twentieth century, these towns were extremely rural and, as seen above, their population densities were noticeably lower than in the present century. Nonetheless, throughout the twentieth century, and especially during the second half of the twentieth and the beginning of the twenty-first century, all the localities supplied by ALJARAFESA experienced continuous population growth. In the first two decades of the twenty-first century, the populations of some more than doubled (Umbrete, Bormujos, Bollullos de la Mitación and Almensilla) while others almost tripled (Espartinas and Palomares del Río) (*La población de Sevilla…* 2007). Some exceptions to this were Carrión de los Céspedes, Isla Mayor, Aznalcóllar, Castilleja del Campo, and Castilleja de la Cuesta, where the population remained stagnant for a range of reasons, including territorial collapse (Castilleja de la Cuesta), the collapse of economic activity (Aznalcóllar), and the road network/distance from the capital (Isla Mayor, Castilleja del Campo). This fast demographic growth from the fourth quarter of the twentieth century onwards started in Mairena del Aljarafe and Tomares in the 1970s, followed by Bormujos, Gines, Castilleja de Guzmán, Palomares, Almensilla, and Espartinas in the 1980s and 1990s (García 2014, 32 and 38–9) and greatly increased the percentages of urban land in the metropolitan area due to the low-density, single-family dwelling housing model adopted. However, this also greatly jeopardised the water-related physical, historical, artistic, and cultural heritage in the district, not only due to the previously-mentioned aggressive urban growth but also because of the destructive processes that affected some unrivalled environments (García and Fernández 2011; García 2013; Cornejo 2007, 189–97; Rosa 2003).[4]

As Table 9.1 clearly shows, the map of municipalities examined in this study has changed markedly since the last third of the last century, as in 1970, almost 90% had fewer than 5,000 inhabitants, whereas in 2021, 77% had over 5,000.

4 One attempt at addressing the loss of identity of dormitory towns in the 1970s was the design of Ciudad Aljarafe, in Mairena del Aljarafe (Navarro 2017).

TABLE 9.1 Localities in the El Aljarafe district (Seville province) distributed by number of inhabitants in 1940, 1970, 2001 and 2021

Inhabitants	1940[a]	1970	2001	2020
Under 1,000	3	2	1	1
1,001–5,000	23	23	11	6
5,001–10,000	3	3	11	12
Over 10,000	0	1	7	11

a The data correspond to only 29 localities as Isla Mayor did not become an individual entity until 1994, when it was separated administratively from the municipality of Puebla del Río.
SOURCE: 1940 CENSO DE POBLACIÓN; 1970 CENSO DE LA POBLACIÓN; 2001 CENSO DE POBLACIÓN Y VIVIENDAS, AND CIFRAS OFICIALES DE POBLACIÓN RESULTANTES… [2021]

In 1970, there was only one locality with over ten thousand inhabitants, while in 2021 there were eleven. Changes in these municipalities have also included their economic diversification (Delgado 2004, 2005, 2006, and 2012; García et al. 2011; García 2014) and deep changes in the district's traditional landscapes (Garcés 1981; Didierjean 1978).

2 The Collapse of the Water Resources

These demographic changes were accompanied by other deeper transformations in water management, which changed from being an individually-managed service in the various municipalities to a collective service that exploited external surface water resources[5] instead of the local, groundwater resources used up to that point, with the connection of individual households to the water supply and the creation of a large sanitation network. Several factors enabled the shift from the classic water supply system that had characterised the El Aljarafe area for centuries: the need to supply water collectively; the implementation of services that would be unviable individually; and bringing down the cost through joint management. The entire process entailed a kind of amnesia regarding the reality of water supply and sanitation that had existed in the district only a few years previously (Navarro 2006, 2007,

5 Yet, when water resources were sought for the Roman town of Itálica, within the municipal boundaries of Santiponce, the water was brought from almost forty km away via an underground/elevated aqueduct (Sánchez and Martínez 2016, 225–8, note 3).

and 2009). It was as if opening up these previously profoundly rural localities to developmentalism marked by intensive urbanisation and new economic activities apart from those related to agriculture was diluting their history and erasing their citizen memory and, with it, their awareness of the long, difficult process of gaining access to quality water and a sanitation system in their own homes (Heller, in press). Some inhabitants might believe that clean water and proper sanitation were available and accessible centuries previously, but this is far from true. Until the mid-twentieth century, the rivers near El Aljarafe, the Guadalquivir and the Guadiamar, guaranteed a water supply for the district, should the need have existed. For normal/everyday use, there were springs and water galleries such as those in Gelves ("Memoria de las actuaciones..." 2021)[6] and, above all, there were endless public and private water fountains and wells. It was these fountains and wells that underpinned the water supply in the district until the requirements for human, agricultural, and industrial needs and equipment grew and, with them, great structures sprang up with systems to collect and distribute water from outside the basins that they supplied, an endeavour in which the ALJARAFESA municipal service company began to excel from the 1970s onwards. Up to this point, the towns and villages in El Aljarafe had been fed by groundwater through the exploitation of, above all, the conveniently-located El Aljarafe aquifer (Hydrogeological unit 05.50; Durán et al. 2003), which, with an area of 350 km^2, could be used for irrigation, cleaning, and animal and human consumption despite the water being very hard and, occasionally, brackish.[7] No less important than the existence of the aquifer was the fact that the entire district could access water at a depth of under fifteen metres, and generally between five and ten metres or even less. However, this arrangement had a major flaw: the sensitivity of the processes to surface pollution, aggravated by the nature of the geological materials in the aquifer (Figures 9.1 and 9.2). As a result of demographic pressure and untreated wastewater from several towns, uncontrolled USW tipping (Durán et al. 2003, 116), and economic activities, especially agricultural activities and the challenges that fertilisers, herbicides, and pesticides presented, contamination was very evident by the mid- twentieth century. The aquifer became contaminated and was overexploited as the overall demand was greater than it could supply, especially in the NW of the district, to the west of the Majalberraque stream,

6 I am grateful to Genaro Álvarez for sending me this initial dissertation on research that is still ongoing.

7 Interviews with Joaquina Román, art teacher, Valencina de la Concepción, 23 February 2006 and Francisco Tirado Cruz, retired water carrier, Carrión de los Céspedes, 9 February 2006.

FIGURE 9.1 Women with their earthenware *cántaros* after collecting water from the *El Rajao* well (Benacazón)
SOURCE: PRIVATE PHOTOGRAPH COLLECTION OF WELL OWNERS

due to shallower saturation levels[8] and the large number of pumps used for irrigation.[9] This was exacerbated by long periods of low recharge and also salinisation (especially at the southern end of the aquifer, in Doñana Natural Park), with water unsuitable for human supply and even of mediocre quality for agricultural use. The Spanish Geological and Mining Institute's analysis of water samples collected at 36 points in the area dates from 1983 (Durán et al. 2003, 116) and, in all likelihood, the water quality has deteriorated since then as compliance with the directives for groundwater management in the El Aljarafe district has been far from strict as far as the reduction of water extraction and the reuse of wastewater are concerned. Notwithstanding, some undeniable milestones have been achieved concerning water savings and its efficient use

8 The shallower saturated and less exploited levels are in the south-western area (Durán et al. 2003, 118).

9 The provisional declaration of the overexploitation of the El Aljarafe aquifer dates from 1988. Around the same time or only a few years later, only 0.4 hm^3 of groundwater per year taken from this aquifer was allocated to the supply of some residential estates in Sanlúcar la Mayor, Almensilla and Bormujos, while between 22 and 25 hm^3 of water was allocated to irrigation. Another 8 hm^3 must be added to this amount due to upwelling or natural lateral overflows, while the total amount of inflows was between 15 and 35 hm^3 per year from rainwater infiltration and, to a lesser extent, returns from irrigation ("Acuífero: Aljarafe…", 4–7; Durán et al. 2003, 115).

FIGURE 9.2 *El Rajao* well (Benacazón). Private wells such as this helped to provide water to residents in the locality
SOURCE: PRIVATE PHOTOGRAPH COLLECTION OF WELL OWNERS

such as the increase in surface resources ("Acuífero: Aljarafe...", 8), which form the basis of ALJARAFESA's current supply.

It was the young women who were responsible for fetching water from wells and fountains, and these became places for socialisation. The so-called *paseo* or walk to collect water was the origin of some symbolic and territorialisation values and also many social relationships with men (Rodríguez 2015), a circumstance that led some fathers in Valencina de la Conception to dig wells in their own homes so that their daughters would not have to go out to collect water from the fountain in Virgen de la Estrella Square.[10] In contrast, the water carriers tended to be men, although there were some cases of female water carriers, such as Sra. Suárez Martínez in Valencina de la Concepción, who used to spend all day carrying water around in her pitcher. Later, José and Francisco Mazo González brought in a mule and a barrel with a tap to carry water. The last water carrier was Enrique Mantos Domínguez in the 1950s.[11] The situation in Carrión de los Céspedes was similar; the last water carrier in the town was Francisco Tirado, whose job it was from the beginning of the 1950s until 1964 when he stopped selling water from the *Pilar* well, the only public supply point in the locality in all those years. Tirado was accompanied by his mule, which pulled a cart carrying a barrel with a tap. He used a trumpet to let people know that he was in the vicinity and used a motor in a cabinet beside the well to bring up the water. It appears that he purchased the business from his predecessor, despite which in 1955 the town council charged him a levy of 4.50 pesetas/day, while the water carrier himself charged 0.35 pesetas (0.35 USD) per jug of water, a price agreed with the Carrión de los Céspedes town council (Figure 9.3). Apart from delivering water to the houses, it seems that every day he took a jug of water to the apothecary to be tested, which to some degree confirmed the quality of the water in the municipality.[12] Along with the use of the services provided by water carriers and public wells, families could have their own wells, which basically depended on their purchasing power. Sometimes people's economic circumstances did not allow them to have their own wells, so shared wells were dug for two, three,[13] and even four houses, making it necessary to go through neighbours' yards.[14] It is estimated that 60% of the houses in Valencina de la Concepción did not have wells, so shared wells were the cheapest solution for less well-off families.[15]

10 Interview with Joaquina Román, art teacher, Valencina de la Concepción, 23 February 2006.

11 Interview with Manuel Delgado, treasurer of Valencina town council, 21 February 2006.

12 Interview with Francisco Tirado Cruz, retired water carrier, Carrión de los Céspedes, 9 February 2006. The journal *Agua y Territorio / Water and Landscape* no. 9 deals with water carriers in Mexico and devotes an entire dossier to the topic coordinated by Evelyn Alfaro-Rodríguez and with the collaboration of lecturers Sánchez-Rodríguez, Rojas-Ramírez, Gómez-Serrano, Rodríguez-Bravo and Rivera-Sánchez.

13 Interview with Francisco Tirado Cruz, "the waterboy", retired water carrier, Carrión de los Céspedes, 9 February 2006.

14 Interview with José Antonio Reinoso Luque, Head of Maintenance, Carrión de los Céspedes town council, Carrión de los Céspedes, 6 February 2006.

15 Interview with Manuel Delgado, treasurer of Valencina town council, 21 February 2006.

FIGURE 9.3 Water carriers in Carrión de los Céspedes, with a specially constructed wooden barrel on a cart drawn by a mule
SOURCE: PRIVATE PHOTOGRAPH SUPPLIED BY FRANCISCO TIRADO

The problems inherent in this supply system have been well documented since the beginning of the twentieth century: water running out during spells of drought (in 1904, for example), the need for public wells which have to be cleaned regularly,[16] etc. As wells are for public use and are normally municipal property, they are to a certain extent subject to health checks as the water is rather unsafe for drinking since it is difficult to chlorinate. Nonetheless, the regular cleaning of municipal wells does not seem to have been as regular as the population's hygiene and health needs required. In 1905, for example, the *Noria La María* well, which supplied most of the residents of Salteras, was extremely dirty and needed to be cleaned due to its improper maintenance over the years, as the local town council readily admitted.[17] In 1908, this well was again repaired, as was the so-called *Concejo* well that supplied other Salteras residents and was also used as a water trough for the livestock in the locality.[18] This recognition of the irregular cleaning of public wells was

16 Salteras Municipal Archive (hereafter SMA), Salteras town council sessions, 2 and 16 June, and 11 August 1904.
17 SMA, Salteras town council session, 18 May, and 1 June 1905.
18 SMA, Salteras town council session, 16 June 1904, and 10 October 1908.

confirmed by the fact that payment for their cleaning was always included in the budget under the entry labelled "unforeseen" or "urgent", which indicated that there were no specific regular allocations for maintaining something as important to the population as the water supply. Despite this, budgetary allocations became increasingly more usual and can be found for cleaning, deepening, and repairing public drinking water fountains in Salteras in 1909,[19] 1910,[20] 1911,[21] 1912,[22] 1916[23] and 1925.[24] Apart from the *Noria* fountain (which included almost 1,200 m^2 of land valued at 300 pesetas) and the *Concejo* fountain (valued at one hundred pesetas), the population could also count on two others sources of safe drinking water to supplement the supply to the locality: the *El Chorrito* fountain (valued at 600 pesetas, with 47,550 m^2 of wasteland used for grazing and with a trough for animals) and the *Alberquilla* fountain (with wasteland valued at 400 pesetas and also with a trough for animals). Lastly, the Salteras town council also possessed another well and water trough called *Fuenblanca*, valued at one hundred pesetas.[25]

The number of public wells in other localities in El Aljarafe may have been similar to the number documented in Salteras. It is reported that in Valencina de la Concepción, in the mid-twentieth century, there was a fountain with a single spout in the current Virgen de la Estrella Square and two wells in gardens (*Pilín* garden and *Manuel* garden) near the locality, on the way to Torrijos farm, with water of variable quality.[26]

3 Factors in the Endogenous Water Resources Crisis

As discussed above, wells used to be the simple solution to water supply in an area with an aquifer quite close to the surface. However, there was also a trade-off to this advantage as the water was easily polluted due to the nature

19 SMA, Salteras town council session, 25 September 1909.
20 SMA, Salteras town council session, 1 August 1910.
21 SMA, Salteras town council session, 4 December 1911.
22 SMA, Salteras town council session, 28 December 1912.
23 SMA, Salteras town council session, 12 October 1916.
24 SMA, Salteras town council session, 8 February 1925.
25 SMA, Salteras town council session, 20 April 1925. Bearing in mind that in 1927 the municipal budget for Salteras did not reach forty thousand pesetas, the expenditure entries were sometimes high, as in 1909, when the repair of the *Alberquilla* and *El Chorrito* wells represented a cost of almost five hundred pesetas.
26 Interviews with Manuel Delgado, treasurer of Valencina de la Concepción town council, 21 February 2006, and Joaquina Román, art teacher, Valencina de la Concepción, 23 February 2006.

of the ground, which is composed of Miocene sandstone and detritus from the Pliocene and Quaternary periods on a non-porous base of blue, sandy loams (in the west, east, and north) and quaternary marshland (in the south) (Durán et al. 2003, 115). Bearing in mind that in 1920 only 42% of the streets in the provincial capital were connected to a sewer network and that the connection rate for dwellings was only 25%, it is easy to draw conclusions about the state of sanitation in the nearby towns and villages in rural El Aljarafe (Matés-Barco 1999, 212–3). Mortality in the Andalusian capital was twice as high as in other Spanish cities due to its large number of cesspits. The existence of these cesspits and water supplied from relatively shallow wells were perfect ingredients for a high incidence of typhoid fever and cholera in El Aljarafe: both the water supply and sanitation systems left much to be desired despite the undeniable advances in bacteriology achieved since the second half of the nineteenth century thanks to Pasteur, Koch (who discovered the cholera bacterium in 1883), Eberth, and Gaffky (who discovered the typhoid fever bacterium between 1880 and 1884) (Ledermann 2003). Although these discoveries were scientific proof of the origin of these two extremely serious water supplies – and sanitation-related diseases –, the relationship between typhoid fever and the consumption of well water contaminated by wastewater had already been known for a century. This knowledge resulted in a fall in the incidence of typhoid fever in Europe from the nineteenth century onward, but the rate was still high in Andalusia until the end of the century and there were sporadic outbreaks during the first half of the twentieth century and even during the second half, which sped up the engineering works to supply drinking water.[27] Water had become one of the main vectors for the spread of disease but was also essential for cleaning and personal hygiene. In this context, Madrid was still experiencing typhoid fever epidemics in 1909 and 1920, while Barcelona suffered almost endemically from the disease, with the worst incidence in years such as 1914. With such a high incidence in the big cities, it is easy to imagine what the situation was in the localities in El Aljarafe, where the problem was a reality in every municipality. In the mid- nineteenth century, typhoid fever was confirmed in localities in the western part of the El Aljarafe district, in places such as Benacazón, Pilas, and Villamanrique (Iglesias 2005, 92). In 1974, Benacazón, for example, was still reporting cases of typhoid fever in the El Prado area, indicating the need for sewers in the El Aljarafe district and the lack of basic sanitation solutions.[28]

---

27 Sanlúcar la Mayor Municipal Archive (hereafter SMMA), ordinary session of Sanlúcar la Mayor town council, 10 July 1947.

28 Benacazón Municipal Archive (hereafter BMA), Benacazón town council extraordinary session, 3 September 1974. The laying of sewers in the El Prado neighbourhood was

However, it was in the 1930s and 1940s that cases of typhoid fever began to surge as a result of water contamination and multiple cases were documented in numerous localities in the district, including Espartinas (1924), Carrión de los Céspedes (1929), Benacazón (1936 and 1940s), Gelves (1941),[29] Sanlúcar la Mayor[30] (1947), and Aznalcázar (1940s).[31] This seems to show that incidence was persistent and outbreaks relatively numerous, although not excessively virulent. All these observations allow it to be confidently stated that there was a divergence between, on the one hand, the scientific discoveries and (post-1886) all the usual legal texts that sought to stop the development of epidemics (Monge 2021; Gómez 2020) and, on the other, the real-world health situation of large sectors of the rural and urban population affected by intense proletarianisation processes. However, the attempt to use State legislation to guarantee water quality and prevent cholera and typhoid fever infections (Matés-Barco 1998) was thwarted by a profusion of sectoral regulations and bylaws that were difficult to implement and clashed with the stark reality of polluted wells caused by uncontrolled waste tipping. Municipal Health Regulations (1925) stated that it was a municipal obligation to supply bacteriologically pure drinking water. However, in addition to the economic difficulties of localities such as these, which were characterised by small populations, there were technical difficulties caused by the use of wells to supply water, with the cesspits (banned under the 1925 Municipal Health Regulations, art. 69) and the fracturing of nearby septic tanks[32] compelling an ongoing search for other wells for drinking water less affected by widespread and invisible contamination.[33] A good

completed in August 1975 with the work contracted out directly as it had been declared urgent (BMA, Benacazón town council ordinary session, 25 August 1975). Even so, between 1970 and 1980, 163 cases were still diagnosed in the Hospital del Mar in Barcelona (Martínez-Roig et al. 2008).

29 Gelves Municipal Archive (hereafter GMA), Gelves town council ordinary session, 15 November 1941.

30 SMMA, Sanlúcar la Mayor town council ordinary session, 10 July 1947.

31 The civil works carried out in this locality at the end of the eighteenth century that resulted in the construction of its outstanding baroque fountain were triggered by the incidence of gastrointestinal infections that caused multiple deaths (García and León 2010, 227).

32 Aznalcázar Municipal Archive (hereafter AMA), Aznalcázar town council ordinary sessions of 3 January, 21 March and 23 May 1903.

33 On this occasion, the typhus vaccination was opted for in Aznalcázar and the water in the public fountain was analysed again after a prudent period of time (AMA, Aznalcázar town council ordinary session of 28 September 1940; García and León 2010, 227 and 235). A few years previously, during the Second Republic, Spain adopted some major sanitation management measures but there had not been enough time for their implementation (Monge 2021, 117–8).

example of this can be found in the locality of Salteras, although similar cases can be found in any other locality in El Aljarafe. At the beginning of the twentieth century there were numerous cesspits spread out throughout the locality. They could be found in public buildings such as boys' and girls' schools and the *Guardia Civil* barracks. It was the town council that had to cover the expenses of both their construction and their cleaning and upkeep because not only did they fill up but, on occasion, they also threatened absolute devastation or collapsed, and the old pit had to be covered up and a new one dug.

It was very difficult to chlorinate the water supplied by wells. Chlorination started to become more widespread in big cities, especially during the 1915–1930 period, but this bore little connection to the reality of the Spanish rural municipalities at the time and for many years after. For a large part of the twentieth century, chlorination was more of an unfulfilled wish for these municipalities than a daily reality, even when houses began to be connected to the water supply.[34] Given that this continued to be the case in the 1970s, it is easy to imagine the deplorable state of the water in the wells in El Aljarafe district at the beginning of the twentieth century. It is important to highlight that not only did abundant lime salts contribute to the hardness of the water, to which the residents seemed to have become accustomed (as typified by the case of Espartinas, for example),[35] but there was also frequent coliform bacillus (E.coli) bacteriological contamination with occasionally severe impacts, as we have just seen.

One of the municipal responsibilities affecting health regulations in the first half of the twentieth century that should be highlighted is that, when the alarm was raised, it was the municipal health inspector and the municipal health boards that were responsible for analysing the water. There is evidence that this was the case in many localities in El Aljarafe such as Aznalcázar, where the Provincial Health Headquarters or Provincial Inspectorate of Health and also the Clinical Analysis Section and the Hygiene Institute used to inspect the wells and intensified the preventative measures since municipalities such as Benacazón and Espartinas had already had persistent levels of water contamination since at least the first third of the twentieth century.[36] This contributed

34 We are preparing an article on this topic on water chlorination in the rural world during the last years of the Franco dictatorship, with special reference to the province of Lleida.

35 Espartinas Municipal Archive (hereafter EMA), book 81, Espartinas Health Board session of 2 June 1924.

36 EMA, book 81, Espartinas Health Board sessions of 27 May 1907 and 2 June 1924. BMA, book 90, Benacazón Local Health Board session and report, 20 July 1940. The Lucas Lavela pharmacy in Seville carried out the bacteriological analyses in 1940 and detected the water contamination in the locality.

to driving up the sense of fear of using well water in both localities and others such as Carrión de los Céspedes and many more. The problem of contamination not only existed in municipal public wells such as the *Pilar* well in Carrión de los Céspedes, the so-called *New Well* in Espartinas and Benacazón's public fountain but also in all the towns' and villages' private wells used by residents, whether they were in their own houses or others. For example, some of the private wells that supplied the townspeople of Benacazón were contaminated between the 1920s and the 1940s. The *El Rajao* well and the fountains at the *Huerta de Don Juan* and the *Huerta del Pino* were contaminated with E.coli, some with extremely high levels of over nine hundred coliform bacilli per litre. Such was the case of the *Huerta de Don Juan* fountain, which led to its use being banned and the water in all the remaining wells being disinfected with products such as *Caporit* (calcium hypochlorite).[37] At least at some point in the century, the responsibility for this fell to the well owners (e.g., at the *El Rajao* well). As far as we know today, in the case of Benacazón, the contamination levels of the public fountain were below those of private wells. This was probably due to the town council, despite its precarious economic situation, being able to act through the municipal health boards,[38] where the issues with water quality had been discussed since the beginning of the twentieth century. This enabled precise and necessary measures to be taken such as the urgent search for pure drinking water[39] when cases of typhoid fever arose. At this point it seemed that the only instruments at hand to solve the problems that were common to the whole of the El Aljarafe district were the vaccination of El Aljarafe inhabitants against typhoid fever (which had begun in republican Barcelona in 1937 and later became the norm in the 1940s) (Monge 2021, 118), isolating the first cases of fever to prevent contagion and boiling water for consumption.

During the whole of the twentieth century, town councils were responsible for the physical protection of the municipal wells and fountains through the erection of railings and iron housings, glass covers and wire fences (an example of this is the Benacazón well[40]), elevating the ground around fountains to prevent water from lying on the surface and seeping into the tank, and sealing the parapets around wells and lining the insides of wells with a coating of cement up to the water level. It is obvious that these measures could do little in a context marked by contamination from black water, waste, etc. that reached

37 BMA, book 90, Benacazón Local Health Board, 23 October 1936.

38 BMA, Benacazón town council extraordinary session, 21 May 1927.

39 EMA, book 81, Espartinas Health Board session, 27 May 1907.

40 AMA, Aznalcázar town council ordinary sessions, 3 January, 21 March and 23 May 1903.

the underground aquifer either through filtration or by way of ropes and buckets, not to mention the detritus that could fall over the parapets, however much care was taken.[41] In Benacazón, for example, it is recorded that by 1909 the railings around the well fountain had been replaced with an iron housing that left only gaps through which the water could be collected. Improvements like these were common in municipalities with this type of supply but probably resulted in only a small increase in the water quality judging by the numerous outbreaks of typhoid fever that occurred for years after.[42] The wells and fountains in all the localities in El Aljarafe would never have been very far from waste, manure, dirty water, stables, pigsties, and places where livestock were raised or kept (Figure 9.4).[43]

The struggle against this unhealthy situation was further exacerbated by the actions taken to find new wells when those that already existed were badly contaminated. Despite their transitory disinfection, it was very difficult to restore the water quality if they were affected by external contamination. The numerous olive oil mills in El Aljarafe from the second half of the eighteenth century and the first half of the nineteenth [44] were contamination blackspots due to the liquid waste produced by olive pressing. This spilt over into the public

41 BMA, Benacazón town council ordinary session, 1 October 1909.

42 BMA, Benacazón town council ordinary sessions, 17 and 24 September 1909 and 9 September 1911.

43 Little study has been done on the fact that in some town councils the existence of rubbish tips on certain plots of land enabled some residents "of recognised probity" to expand their properties by incorporating these plots into their properties with the promise of disposing of the rubbish. Such is the case of a 40 m^2 plot in Salteras in 1904 (SMA, Salteras town council sessions, 3 and 10 March 1904).

44 Until the nineteenth century, the landscape of El Aljarafe was mainly characterised by Mediterranean crops that thrived in the high-quality land and the micro-climate conditions of an ocean-influenced Mediterranean climate, not to mention the abundance of sunshine, mild winters, and seasonal rainfall (Borrero 1983, 57–116 and 217–43; Carabaza 1999). Cereal cultivation stood out historically, especially wheat and barley. Much less land was given over to grapevines and olive trees, although during Roman times olives may have been important, as they were during the rise in agriculture in the district from the fifteenth century on (Borrero 1982; González 2015; Didierjean 1978, 25–30). Some municipalities, including Benacazón and Carrión de los Céspedes, for example, may have grown more olives than others and this was also quite a prominent crop in Huévar, Pilas and Palomares, but this does not mean that the towns and villages in El Aljarafe were in any way lacking in olive mills. For example, in the second half of the eighteenth century, Aznalcázar had four olive mills, as well as three flour mills and three winepresses, while the now abandoned Gelo farmstead was devoted to olive growing during the Late Middle Ages, with the assignation of seigneurial dues for the use of the mill in nearby Benacazón (González 1982; Prieto 2005; Hernández 2005; Iglesias 2005, 95; Herrera 2010, 31; López 2012; Valiente and Martínez 2005). Apart from the traditional beam mills, there

FIGURE 9.4 Eighteenth-century baroque fountain in Aznalcázar. An important example of the El Aljarafe district's water heritage
SOURCE: PHOTOGRAPHED BY THE AUTHOR

highways to the detriment of the groundwater, passing people, and animals. The mills in Daoiz Street and Don Pelayo Street in Salteras were notorious. An added aggravating circumstance was that neither of their owners, Enrique González Navarro and Antonio de los Reyes Jiménez, paid the local industrial tax.[45] A similar situation existed regarding old cemeteries. After a long process that began in 1908,[46] the old cemetery in Salteras (called the "Jesus Cemetery")

were also hydraulic mills in the Late Middle Ages, such as those studied by Valor Piechotta (1989) on the Riopudio and Majalberraque rivers.

45 SMA, Salteras town council ordinary sessions, 7 December 1907 and 18 December 1909. Threats of proceedings and fines did not seem to have much effect as two years later Antonio Reyes was still being asked to dispose of the liquid waste, to not allow it to stagnate, and to ensure that it was covered when it was transported through the town streets to the countryside.

46 SMA, Salteras town council ordinary sessions, 6 June and 18 July 1908. Despite the urgency of its construction for health reasons, due to which the work was not tendered out, the cemetery's construction took an inordinate number of years and was paid for out of the town coffers (SMA, Salteras town council sessions, 2 January and 4 December 1909, 23 November 1912, 4 and 11 January 1913).

was replaced with a new cemetery that would take the name of San Carlos and was built in 1921. The old cemetery was within the town itself and as it had been impossible to carry out any more burials there; the General Benevolence and Health Directorate had ordered its closure in 1908. However, it continued to operate despite being a constant public health risk. The old graveyard was not cleared until 1926 when all the human remains were removed and taken to the new cemetery.[47] It was precisely around this same time that health and hygiene started to take on greater prominence in Salteras' budgets, as did public works (10% of the budget),[48] with specific budgetary allocations for street cleaning, the removal of manure, etc. and for paving some of the streets in the centre of town, including González Eiris, Nuestra Señora de la Oliva and General Queipo de Llano streets (1942), where cobblestones were laid on a bed of sand,[49] and Daoíz, Cervantes and José Cortés streets (1966), with cobblestones and mortar and a 15 cm concrete base[50] and pavements made out of concrete slabs with sumps.

There were other health threats apart from the old cemeteries that, as we have seen, had obvious health effects, even in towns with moderate growth such as Salteras (1594 inhabitants in 1900 and 1898 inhabitants in 1930). One threat that adversely impacted public health was the slaughtering of livestock, which was done on the properties of the butchers who, generally, lived in the town centres and, as is logical, did not provide any healthy hygienic conditions as the bins in which the offal and manure were collected were in their yards.[51] Consequently, plans to build public slaughterhouses on the outskirts of the town began to proliferate in health legislation and regulations.

Urban solid waste was also becoming a serious problem by the end of the 1970s to the extent that, in a draft project that he prepared for the treatment of urban solid waste in Seville province, the Provincial Government's industrial engineer, Francisco Llatas Duque, mentioned that "the water and the air accept everything that is thrown at them, although their deterioration can be

47 Salteras' new town cemetery was valued at 12,000 pesetas, probably the cost price of its construction a few years previously (SMA, Salteras town council ordinary session, 20 April 1925).

48 SMA, Salteras town council ordinary sessions, 9 November 1923, and 1 February 1924.

49 Project to pave González Eiris, Nuestra Señora de la Oliva and General Queipo de Llano streets, 1942 (SMA, 164).

50 Project to pave José Cortés Street, 1966 (SMA, 164, and Salteras town council ordinary sessions, 29 February 1924, 26 November 1926, and 2 December 1927).

51 SMA, Salteras town council ordinary session, 6 June 1908.

considerable and irreversible".[52] Waste was already observed to have increased substantially by weight due to consumerism and the higher standard of living, especially in the most populated towns in the El Aljarafe district such as Mairena del Aljarafe (11,705 inhabitants), La Algaba (11,655), Pilas (9,448), Castilleja de la Cuesta (8,862) and Sanlúcar la Mayor (7,238). Waste volumes had also increased due to a decrease in organic waste being offset by an increase in non-biodegradable paper, cardboard, plastics, and containers. Each of the abovementioned localities, the five largest in the El Aljarafe district, produced over 150 t/month at the time, with La Algaba generating as much as 230 t/month. The total for all the localities that today come under ALJARAFESA was 2467 t/month. It was not only the volume that was worrying; the procedure used for treating the waste was also a cause for concern. Practically half the waste generated in the El Aljarafe district was sent directly to landfill without any treatment at all. This had serious negative consequences for the environment and the groundwater, with a parallel rise in unpleasant smells, fumes, insects, and rodents. Only 25% of waste disposal was controlled and, despite being safer and having a lower operating cost at the end of the 1970s, it was still not completely devoid of health risks.

4 Sources

Aznalcázar Municipal Archive (AMA).
Benacazón Municipal Archive (BMA).
Espartinas Municipal Archive (EMA).
Gelves Municipal Archive (GMA).
Salteras Municipal Archive (SMA).
Sanlúcar la Mayor Municipal Archive (SMMA).

References

"Acuífero: Aljarafe. Unidad hidrogeológica: Aljarafe (05.50)." In *Catálogo de acuíferos con problemas de sobreexplotación o salinización.*

Borrero Fernández, M. 1982. "Ordenanzas del Aljarafe (Siglo XVI)." *Historia. Instituciones. Documentos* 9: 425–52. https://doi.org/10.12795/hid.1982.i09.09.

52 Francisco Llatas Duque: "Anteproyecto de tratamiento de residuos sólidos urbanos en la provincia de Sevilla". Excma. Diputación Provincial de Sevilla, Seville, January 1979. SMA, 165.

Borrero Fernández, M. 1983. *El mundo rural sevillano en el siglo XV: Aljarafe y Ribera*. Sevilla: Diputación Provincial de Sevilla.

Borrero Fernández, M. 2005. "Aljarafe y Ribera del Guadalquivir en la Edad Media." In *Actas II Jornadas de Historia sobre la provincia de Sevilla. Aljarafe-Marismas*, 57–75. Ronda: Editorial La Serranía-Asociación Provincial Sevillana de Cronistas e Investigadores Locales. Accessed 5 January 2023. https://dialnet.unirioja.es/servlet/libro?codigo=355530.

Carabaza Bravo, J. M. 1999. "El paisaje agrícola del Aljarafe sevillano." *Revista del Instituto Egipcio de Estudios Islámicos* 31: 225–42.

"Censo de población de 1940 (Hecho). Clasificación por municipios. Provincia de Sevilla". Accessed 5 January 2023. https://www.ine.es/inebaseweb/pdfDispacher.do?td=120587&ext=.pdf.

"Censo de la población de 1970. Provincia de Sevilla. Relación de las poblaciones de Derecho y de Hecho por municipios". Accessed 5 January 2023. https://www.ine.es/inebaseweb/pdfDispacher.do?td=144411&ext=.pdf.

"Censo de población y viviendas 2001. Provincia de Sevilla". Accessed 5 January 2023. https://www.juntadeandalucia.es/institutodeestadisticaycartografia/censos/censo2001/Fichas/CSIND41.htm.

"Cifras oficiales de población resultantes de la revisión del Padrón municipal a 1 de enero [2021]". Accessed 5 January 2023. https://www.ine.es/jaxiT3/Datos.htm?t=2895.

Cornejo Ortiz, S. 2007. "Del conocimiento a la protección del patrimonio en pequeños municipios: el Aljarafe sevillano." PhD diss., University of Seville.

Delgado Bujalance, B. 2004. *Cambio de paisaje en el Aljarafe durante la segunda mitad del siglo XX*. Seville: Diputación Provincial de Sevilla.

Delgado Bujalance, B. 2005. "Actores y factores de la configuración del paisaje del Aljarafe Sevillano en un contexto metropolitano: una aproximación desde la ecología política." *Xeografica: revista de xeografia, territorio e medio ambiente* 5: 19–41.

Delgado Bujalance, B. 2006. "Transformaciones rápidas en los paisajes metropolitanos del Aljarafe sevillano." *Ería* 70: 161–73.

Delgado Bujalance, B. 2012. "El cambio de paisaje en la cornisa oriental del Aljarafe (Sevilla)." PhD diss., University of Seville.

Didierjean, F. 1978. "Le paysage rural Antique au Nord-Ouest de Séville: Campo et Aljarafe." *Mélanges de la Casa de Velázquez* 14: 5–33. https://doi.org/10.3406/casa.1978.2262.

Durán Valsero, J. J. et al. (eds.). 2003. *Atlas hidrogeológico de la provincia de Sevilla*. Madrid: Instituto Geológico y Minero de España-Diputación Provincial de Sevilla.

El Faiz, M. 1991. "Aljarafe de Sevilla: un jardín d'essais pour les agronomes de l'Espagne musulmane." *Hesperis Tamuda* 29 (1): 5–26.

Garcés Olmedo, A. 1981. "Elementos para el estudio del paisaje rural del Aljarafe según las "Respuestas Generales" (1751)." *Archivo Hispalense* 64 (193–194): 189–212.

García Brenes, M. D., and R. Fernández Cañero. 2011. "Ruptura de la sostenibilidad en la comarca del Aljarafe (Sevilla). La dialéctica olivar-urbanización." *Scripta Nova* 15, 348–86.

García García, A., V. Fernández Salinas, and L. Andrés Zambrana. 2011. "Luces y sombras de los espacios públicos del Aljarafe sevillano." *eDap: documentos de arquitectura y patrimonio* 3–4: 120–7.

García Martín, M. 2013. "Percepciones y valoraciones sociales del territorio en las aglomeraciones urbanas: paisaje y lugar en el Aljarafe (Sevilla)." PhD diss., University of Seville.

García Martín, M. 2014. "Transformaciones territoriales recientes en el Aljarafe sevillano: de la vocación rural a la integración metropolitana." *Cuadernos Geográficos* 53 (2): 25–53.

García Martín, M. 2016. "Representaciones artísticas de la ciudad de Sevilla desde la cornisa del Aljarafe." In *Ciudad y Comunicación*, edited by M. A. Chaves Martín, 301–10. Madrid: Universidad Complutense de Madrid.

García Parra, P. J., and D. A. León García. 2010. "La fuente vieja de Aznalcázar. El Santuario del Agua." In *Actas VII Jornadas de Historia sobre la provincia de Sevilla. El Aljarafe Barroco*, edited by J. A. Filter Rodríguez, 227–41. Seville: Asociación Provincial Sevillana de Cronistas e Investigadores Locales. Accessed 5 January 2023. https://dialnet.unirioja.es/servlet/libro?codigo=707822.

Gómez Martínez, M. R. 2020. "El binomio miseria-enfermedad: un estudio socio histórico y legislativo sobre las fiebres tifoideas en España (1886–1930) desde la Historia de la Salud." *Cultura de los Cuidados* 24 (56): 101–14. https://doi.org/10.14198/cuid.2020.56.07.

González Arce, J. D. 2015. "La producción oleícola del Aljarafe según el diezmo del almojarifazgo de Sevilla (siglo XV)." *Historia Agraria* 65: 43–74.

González Jiménez, M. 1982. "Nuevos datos sobre la repoblación del Aljarafe sevillano durante el siglo XIV." In *En la España medieval. Estudios en memoria del Profesor D. Salvador de Moxó*, edited by M. A. Ladero Quesada, 425–34. Madrid: Universidad Complutense de Madrid.

Heller, L. (in press). *Los derechos humanos al agua y al saneamiento*. Jaen: Universidad de Jaén.

Hernández González, S. 2005. "La ermita de Gelo (Benacazón): noticias documentales sobre su historia y patrimonio artístico". In *Actas II Jornadas de Historia sobre la provincia de Sevilla. Aljarafe-Marismas*, 257–69. Ronda: Editorial La Serranía-Asociación Provincial Sevillana de Cronistas e Investigadores Locales. Accessed 5 January 2023. https://dialnet.unirioja.es/servlet/libro?codigo=355530.

Herrera García, A. 1981. *El Aljarafe durante el Antiguo Régimen*. Seville: Diputación Provincial de Sevilla.

Herrera García, A. 2010. "Panorama sociopolítico del Aljarafe sevillano en la época del Barroco." In *Actas VII Jornadas de Historia sobre la provincia de Sevilla. El Aljarafe

Barroco, edited by J. A. Filter Rodríguez, 21–33. Seville: Asociación Provincial Sevillana de Cronistas e Investigadores Locales. Accessed 5 January 2023. https://dialnet.unirioja.es/servlet/libro?codigo=707822.

Iglesias Rodríguez, J. J. 2005. "La comarca del Aljarafe-Marismas a fines del Antiguo Régimen." In *Actas II Jornadas de Historia sobre la provincia de Sevilla. Aljarafe-Marismas*, 77–118. Ronda: Editorial La Serranía-Asociación Provincial Sevillana de Cronistas e Investigadores Locales. Accessed 5 January 2023. https://dialnet.unirioja.es/servlet/libro?codigo=355530.

Ledermann D. W. 2003. "Una historia del bacilo de Eberth desde Junker hasta Germanier." *Revista chilena de infectología* 20: 58–61. https://doi.org/10.4067/S0716-10182003020200020.

López Sánchez, J. P. 2012. "Las canciones de trilla en el Aljarafe sevillano." *Olivar* 13 (18): 277–294.

Martín Ponce, A. M. 2010. "El Aljarafe y su patrimonio cultural: las haciendas." *Ben Baso* 21: 35–37.

Martínez-Roig, A., I. Morales, R. Berrueco *et al.* 2008. "Retrobament amb la febre tifoide a Catalunya al segle XXI. Comentaris clínics i al tractament." *Pediatria Catalana* 68: 12–18.

Matés-Barco, J. M. 1998. *Cambio institucional y servicios municipales: una historia del servicio público de abastecimiento de agua*. Granada: Comares.

Matés-Barco, J. M. 1999. *La conquista del agua*. Jaen: Universidad de Jaén.

"Memoria de las actuaciones de espeleología aplicada en galerías artificiales de Gelves (Sevilla)". 2021. Sociedad Espeleológica Geos.

Monge Juárez, M. 2021. "La II República contra las fiebres tifoideas: vertebración del territorio y agua potable. Aproximación histórica a las políticas legislativas y preventivas, 1931–1939." *Ebre38* 11: 111–23.

Navarro de Pablos, F. J. 2017. "La utopía de Ciudad Aljarafe: identidad vecinal frente a la ciudad dormitorio." In *Periferias urbanas. La regeneración integral de barriadas residenciales obsoletas*, edited by M. T. Pérez Cano, and D. Navas Carrillo, 67–81. Seville: Universidad de Sevilla.

Navarro García, J. R. 2006. "Agua y enfermedad en el Aljarafe durante la crisis del sistema de abastecimiento clásico (1900–1950)." In *El debate del agua en Jalisco y Andalucía*, edited by J. R. Navarro García, and J. Regalado Santillán, 95–152. Seville: Consejería de Cultura, Junta de Andalucía.

Navarro García, J. R. 2007. *Entre el recuerdo y la Historia. Agua y salud en el Aljarafe en la primera mitad del siglo XX*. Seville: Aljarafesa.

Navarro García, J. R. 2009. "El saneamiento en la comarca sevillana del Aljarafe en la primera mitad del siglo XX." *Anduli* 8: 99–120.

Ordóñez Agulla, S., and J. Beltrán Fortes. 2005. "Panorama de la romanización en la comarca del Aljarafe." In *Actas II Jornadas de Historia sobre la provincia de Sevilla.*

Aljarafe-Marismas, 21–55. Ronda: Editorial La Serranía-Asociación Provincial Sevillana de Cronistas e Investigadores Locales. Accessed 5 January 2023. https://dialnet.unirioja.es/servlet/libro?codigo=355530.

La población de Sevilla. Datos básicos de Sevilla y su relación con Andalucía y España. 2007. Cuadernos Fundación BBVA. Accessed 5 January 2023. https://www.fbbva.es/wp-content/uploads/2017/05/dat/cuaderno_sevilla.pdf.

Prieto Pérez, J. O. 2005. "Aznalcázar en la segunda mitad del siglo XVIII." In *Actas II Jornadas de Historia sobre la provincia de Sevilla. Aljarafe-Marismas*, 143–59. Ronda: Editorial La Serranía-Asociación Provincial Sevillana de Cronistas e Investigadores Locales. Accessed 5 January 2023. https://dialnet.unirioja.es/servlet/libro?codigo=355530.

Rodríguez Becerra, S. 2019. "Símbolos y rituales religiosos en el Aljarafe sevillano." *Antropología Experimental* 19: 135–52. https://doi.org/10.17561/rae.v19.13.

Rodríguez Fernández, J. 2015. "Agua, poder, sociabilidad y desigualdades de género en las fuentes públicas de las villas alavesas (1450–1550)." In *La percepción del agua en la Edad Media*, edited by M. I. del Val Valdivieso, 17–37. Alicante: Universitat d'Alacant.

Rosa Jiménez, C. J. 2003. "Transformaciones metropolitanas en el Territorio Cultural del Aljarafe." PhD diss., University of Seville.

Sánchez López, E., and J. Martínez Jiménez. 2016. *Los acueductos de Hispania. Construcción y abandono*. Madrid: Fundación Juanelo Turriano.

Tahiri, A. 2011. "El Aljarafe, Corona de Sevilla." *eDap: documentos de arquitectura y patrimonio* 3–4: 114–8.

Valiente Romero, A., and I. Martínez Laguna. 2005. "Sociedad y cultivo de la vid en el Aljarafe sevillano: el caso de Valencina de la Concepción." In *Actas II Jornadas de Historia sobre la provincia de Sevilla. Aljarafe-Marismas*, 213–22. Ronda: Editorial La Serranía-Asociación Provincial Sevillana de Cronistas e Investigadores Locales. Accessed 5 January 2023. https://dialnet.unirioja.es/servlet/libro?codigo=355530.

Valor Piechotta, M. 1989. "Molinos hidráulicos de rodezno en el Aljarafe sevillano." In *El agua en zonas áridas. Arqueología e Historia. Hidráulica tradicional de la provincia de Almería*, edited by L. Cara Barrionuevo, 737–52. Almeria: Instituto de Estudios Almerienses.

Zurita Chacón, M. 2017. "Señoríos, órdenes militares y religiosas en el territorio Aljarafe-Doñana.". In *Actas XIII Jornadas de Historia y Patrimonio sobre la provincia de Sevilla. Las órdenes religiosas y militares en la provincia de Sevilla* (*siglos XIII-XX*), edited by J. A. Filter Rodríguez, 159–176. Seville: Asociación Provincial Sevillana de Cronistas e Investigadores Locales. Accessed 5 January 2023. https://dialnet.unirioja.es/servlet/articulo?codigo=6726636.

CHAPTER 10

The “Price” of Water in Andalusia: An Approach to the Tariff Systems of the Domestic Water Supply Service

María José Vargas-Machuca Salido

Abstract

From a legal point of view, in Spain, water as a good has no price, as it is part of the public domain. What domestic consumers pay for the water they consume is the consideration for its supply and treatment services. The competence regarding the setting of these remunerations falls on the local entities whose legal nature will determine the legal form of these considerations. Following the Water Framework Directive 2000/60/EC, water tariffs have become an important tool to achieve economic efficiency, environmental sustainability, and social equity. In this work, once the competence framework for the approval and setting of tariffs in Andalusia has been delimited, a comparative approximation of the tariff systems of the domestic water supply service in in the eight provincial capitals of this region is carried out.

Keywords

Andalusia – domestic water supply – integral water cycle – Water Framework Directive – water tariffs

1 Introduction

Over the coming years, water management will face major challenges derived from the increase in demand for consumption, industrial use and energy generation and also due to the climate conditions suffered by the planet. This scenario requires a comprehensive management of water resources and the design of tariff policies that are aligned with these challenges.

As a scarce and valuable resource, water requires an efficient and sustainable management, both from an economic and environmental point of view. In turn, as it is an essential good for life, it is necessary to guarantee its universal

 | DOI:10.1163/9789004679634_011

access so that the whole population can enjoy and afford a minimum level of water and sanitation services.

These factors condition the design of water cycle management and convert the tariff structure into a key tool for achieving economic efficiency, environmental sustainability and social equity. In the case of the European Union, these objectives are laid out in the Water Framework Directive 2000/60/EC, which establishes a framework for community action in the field of water resource management and applies the cost recovery principle and the criterion of who pollutes pays.

From a legal point of view, in Spain, water as a good has no price because it forms part of the public domain. Domestic consumers pay for water supply and treatment services. The idea is that water pricing policies can provide users with enough incentive to use water resources efficiently and, therefore, contribute to the environmental objectives.

The purpose of this study is to analyse the tariff structures applied in the eight provincial capitals of Andalusia, focusing on the supply of water for domestic use.

To do this, following this introduction, the chapter is divided into four parts and a conclusions section. First, a brief description is made of the water cycle, indicating who is responsible for the management in each of its phases. Next, the objectives that water resource management should fulfil and their materialisation in the Water Framework Directive are analysed. Then, from a theoretical point of view, the different forms of tariff that can be used to determine the prices of urban water are reviewed. Finally, the tariff structures applied in the capitals of the Andalusian provinces are studied. The study concludes with a section that lists the bibliographical references used and an annex containing a breakdown of the tariffs in blocks of consumption for the different cities analysed.

2 The Integral Water Cycle

Water is an essential good which is also scarce. It is, therefore, necessary to establish infrastructures that guarantee its safe and efficient supply. This requires elements for its storage, purification, distribution and treatment that are integrally managed. This series of infrastructures and their management system is called the "integral water cycle" (PWC 2018, 7). This cycle can be divided into two levels: the supply and sanitation of water. In turn, each of these includes different phases (CNMC 2020, 14–5; PWC 2018, 7–8).

Supply covers the activities of capture, transport, purification, and distribution. In a first phase (supply of "upstream water"), the water is captured from different sources such as reservoirs, rivers, wells, and the sea (desalination). It is stored and transported to the drinking water treatment stations for its purification and, from there, to the municipalities. Agricultural consumption and that of large industries or electrical power plants correspond to this level as these destinations do not require the purification of the water. In a second phase ("downstream water"), once apt for human consumption, the water is transported to the outlets and meters of buildings and homes.

Water sanitation also includes two phases. First, the wastewater is collected and transported to the main sewage drains (wastewater collection). Then, the water is conveyed to the wastewater treatment plants where, after its treatment, it is converted into clean water apt to be returned to the natural environment (treated wastewater discharge).

2.1 *The Management of Urban Water Services*

Different administrative levels intervene throughout the integral water cycle: the State, the regional governments, and the municipalities, which assume the competencies attributed to them by the Spanish Constitution, the different autonomy statutes, and Law 7/1985 of 2 April, regulating local government rules (LRBRL). In this respect, the management and use of water resources are the responsibility of the State and regional governments and the criterion for distributing the competencies is the territory through which the water runs.

Meanwhile, the provision of urban water supply and sanitation services ("urban water cycle") corresponds to the municipalities. For these services, the local entities enjoy a reservation of activity (art. 86.2 LRBRL). However, in those municipalities with a population of under 20,000 inhabitants, the provincial government (or equivalent body) coordinates the supply of drinking water to homes and the collection and treatment of wastewater, with the agreement of the affected municipalities. This provision is carried out directly by the provincial government or through consortia, associations, or other formula, except when the municipality justifies the possibility of providing these services at a lower cost (art. 26.2 LRBRL).

Public intervention has been traditionally justified by different factors. First, the consideration of water as a good necessary for life generates positive externalities. Therefore, guaranteeing universal access to clean water is both an individual right and also beneficial for the whole of society. Second, public intervention ensures the correct allocation, protection, and conservation of a resource whose ownership rights are not clearly defined as it is a common good. Finally, the introduction of competition in the necessary network provision of this good would give rise to the duplication of costs and inefficient solutions, so the existence of a sole provider in the form of a natural monopoly would be the best solution. Taking these factors into account, the direct management by local entities (local councils) would be the best way of guaranteeing the universality of the service in healthy conditions, avoiding the abuse of market dominance (González 2005, 306–7).

The LRBRL establishes that the management of the local public services does not always have be conducted internally by the municipalities authorities and the supply and sanitation services can be carried out under direct or indirect forms of management. In the first case, the management corresponds to the local entity itself, to another autonomous body, a public business entity or a commercial company with publicly owned capital, all on a local scale. The creation of these new municipal institutions decentralises the activities and decongests the undertakings of the local governments. Although their constitution involves a certain level of autonomy of action, the local governments maintain direct control of important decisions. The second form, indirect management, can be carried out through service concession contracts, interested management, concerted management, leasing or through mixed companies with capital only partly publicly owned (CNMC 2020, 17; González 2005, 307).

Some of the indirect forms of management require the total privatisation of the services. The necessary network provision of these services, with significant economies of density, implies a shift from a public monopoly to a private one, which necessarily requires the supervision of the public authorities. The obligations of the private body must be very clear at all times and stipulated and listed in the solicitation documents.

In addition to decongesting the activity of the local governments, the participation of the private sector in the management of the urban water cycle is also considered to contribute to reducing the costs of the service. On the one hand, due to the possible competition between companies wishing to manage the service in the corresponding tender and on the other hand, due to the supposed greater efficiency attributed to this type of management as opposed to the public initiative. Finally, if the private company has experience in the sector it can contribute its know how (González 2005, 307).

Independently of the debate on whether private management enables greater efficiency than public management, the reality is that over the last few years there has been a trend towards privatisation on an international scale in water supply and sanitation services. In the case of Spain, in 2020, thirty-five per cent of the management of supply was covered by public companies, 33% by private companies, 22% by mixed companies and 10% by the municipalities (AEAS and AGA 2020).

3 The Objectives of Water Resource Management: The Water Framework Directive

In the coming years, water management will face major challenges derived from the increase in the demand for water due to the exponential growth in the population and the new requirements for industrial uses and energy generation. This scenario requires an integrated management of water resources which, among other factors, conditions the tariff policies applied. The most important challenge involved in establishing tariffs in the supply of water is reconciling the different objectives, which are sometimes in conflict with each other. These many objectives can be structured around four dimensions (OCDE 2009, 80–81):

- Environmental sustainability: as a scarce natural resource, water should be used taking into account the characteristics of the natural capital in order to preserve it for future generations. Saving water forms part of this objective, which requires its efficient use, that is, avoiding wasteful practices that unnecessarily pressure the availability of water and the design of suitable mechanisms for its reuse.
- Economic efficiency: as a valuable economic good, water should be allocated to the uses that maximise overall benefits for society (allocative efficiency).
- Financial sustainability: the cost recovery principle establishes the need to cover the total cost in all water uses, enabling the financial sustainability of the investments and the services of water supply. The management of water resources and the provision of supply and sanitation services are activities that require investment in expensive infrastructures, which, in turn, require income in order to maintain them over time. In this respect, it is not only the level of the tariffs that are important, but the transparency and stability of the price setting process.
- Social interest: the consideration of water as an essential good for life means that its universal access must be guaranteed. The entire population, including the low-income groups, should be able to access and afford a minimum level of water supply and sanitation services. The regulators should prevent the suppliers, who generally operate under a natural monopoly regime, from establishing a tariff system that exploits this position in terms of excessive profits or a reduction in the operating efficiency. Attention should be focused mainly on how to protect vulnerable groups to guarantee their access to water services and to ensure that that they can continue to do so in an affordable way over time. Within this context, the average tariff level

is not as important as the way in which the costs are distributed among the different groups through the tariff structures.

In Europe, these principles constitute the basis of the Water Framework Directive[1] which came into force in the year 2000 and which establishes a community action framework with respect to the management of water resources. Article 9 indicates that "Member States shall take account of the principle of recovery of the costs of water services, including environmental and resource costs having regard to the economic analysis conducted [...], and in accordance in particular with the polluter pays principle". Member states should ensure "that water-pricing policies provide adequate incentives for users to use water resources efficiently, and thereby contribute to the environmental objectives" of the Directive. We can say, therefore, that the Water Framework Directive establishes two fundamental objectives for tariff structures: to favour cost recovery and influence the consumption habits of citizens by promoting saving and the rational use of this scarce resource (Daza 2008, 29). The difficulty that often arises is the alignment of all of these objectives in a single tariff instrument. This has led to the emergence of a wide variety of pricing systems and tariff structures that have sought to resolve the conflicts between these objectives (Suárez-Varela 2020, 207).

4 The "Price" of Urban Water: Tariff Systems

From a legal point of view, in Spain, water as a good has no price because it forms part of the public domain. Domestic consumers pay for the supply and treatment services of water. This amount is made up of two components (CNMC 2020, 21):

- The compensation for the supply and sanitation services of urban water which are set on a local level and which, in some cases, are reviewed on a regional level.
- The state and regional charges and fees that cover the cost of the water infrastructures, to achieve certain environmental goals or tax the use of the public water domain. These elements are not borne directly by the final users, given that the passive subject is the concession holder. In the case of the urban water cycle, the local entities charge the consumers these fees in the invoice of the water supply service.

1 Directive 2000/60/EC of the European Parliament and Council of 23 October 2000, establishing a community action framework with respect to water policy.

The destinations of the water in the urban environment include domestic uses corresponding to private homes (72 per cent), industry and construction uses (11 per cent), municipal consumption (9 per cent) and tourism services and uses (8 per cent) (INE 2018).

This study will focus solely on the "prices" paid by the domestic consumers for the supply of urban water which, in some cases, are added to the invoices of the afore-mentioned charges.

4.1 *Prices, Tariffs, and Charges of the Urban Water Supply Services*

Determining the amounts that have to be paid by the users for the urban water services is the responsibility of the local entities. The legal nature of these payments depends, in turn, on the legal status of the agent providing the service.

From 2017, in the case where the services are provided by the local entities, the amount is a tax, set in accordance with the Royal Legislative Decree 2/2004 of 5 March.[2] However, if a company (public, private or mixed) provides the services, the amount for these services is considered as a "provision of a public untaxed asset".[3] In the case of supply services, these amounts are subject to authorised prices which implies that they are subject to a double approval system: the "prices" set by the managers of the services should be authorised first on a local level and, subsequently, by the regional government through the body responsible for authorised prices. The objective of this double authorisation is to verify that the amounts proposed correspond to the characteristics of the services and the possible variations in the costs of the supplying agents (CNMC 2020, 22–3).

In addition, there are other taxes and charges set by the state or regional authorities and other bodies related to the supply of water and which the local entities usually transfer to the users through water bills. This is the case, on a state level, of the charge for using the public water domain, the dumping fee or the regulation charge and tariff for using water. On the other hand, the Autonomous Regions also have these kinds of taxes related to the water cycle, usually with environmental objectives or for financing, exploiting and maintaining the water infrastructures, as in the case of Andalusia (CNMC 2020, 23–4).

2 Royal decree approving the Restated text of the Law Regulating Local Finances (TRLHL).

3 Article 20.6 of the TRLHL introduced by the 12th Final Provision of Law 9/2017 of 8 November regarding Public Sector Contracts, which transposes into the Spanish national legal order the Directives of the European Parliament and Council 2014/23/EU and 2014/24/EU of 26 February 2014.

4.2 *Tariff Types: A Theoretical Approach*

The charging for services corresponding to the urban water cycle is the formula established as compensation for the supplying body so it may achieve cost recovery and an economic-financial balance of the service. Irrespective of its legal form as a rate or public price, the compensation for the water supply services can take the form of different types of tariffs. In general, the tariffs for the supply of water usually applied are the result of a combination of some (or all) of these elements (Álvarez, García, and Suárez 2001, 17–8):

- A fixed connection fee, which can be paid when the connection is made or periodically together with the payment for consumption.
- A fixed amount for accessing the supply systems, unrelated to consumption, which is usually established in accordance with the characteristics of the consumer.
- A variable amount applied to the volume of water consumed.
- A minimum quota responding to the minimum volume of the service, which must be paid in each period even though this level of consumption is not reached.
- Block tariffs, which imply the application of different rates, increasing or decreasing, for different consumption intervals.

The varied combinations of these elements give rise to a wide range of tariff structures which we can summarise as follows (Babiano 2020, 18–9; Instituto Aragonés del Agua 2013, 19–25):

- Fixed fee tariff: a fixed fee is established based on average data determined statistically, but without taking into account the real consumption of water. The fees can be the same for all users or vary depending on the type of activity of the building (domestic, restaurant, industry, etc.).
- Unit tariff (or monomial): comprises solely of one variable part which includes a price per cubic metre consumed.
- Binomial tariff: divided into two components (fixed and variable) which seek to adapt the revenue to the cost structure of the water services.
  - Fixed or service fee: the amount is usually determined by the calibre of the meter, with higher costs for those connections with larger diameters.
  - Variable or consumption fee: the part paid by the users in accordance with their consumption and which can be defined in different ways.
    - Constant fee: within the same water use (domestic, commercial and industrial) all the consumption is invoiced at the same price.
    - Block tariffs: consumption is broken down into predetermined thresholds to which different prices are applied. These prices can be decreasing (currently in disuse) or increasing. The latter also include tariffs related to the number of members in the household, whereby

the tariff is progressive as it calculates the individual consumption of each person in the home.

 - Minimum tariffs: the minimum amount of water billed, irrespective of whether it is actually consumed or not. Above this minimum amount, the water is invoiced depending on the consumption with a constant price or according to a block tariff system. This could be considered as a binomial tariff as the fixed fee includes the minimum consumption and the first consumption block has a price of zero.

- Other tariff systems: those that establish different prices depending on different criteria: different areas of cities, the cadastral value of the homes, season of the year, time segments, etc. These systems are less frequently used than those previously described.

Finally, it is worth mentioning that, in some cases, special tariffs may be approved for households with particular circumstances. This is the case, for example, of the tariffs for large families or for low-income households, with the setting of differentiated prices for these groups (retired people, young people, the unemployed...). In these cases, a minimum consumption is usually established which is considered as a basic need for which an affordable price is set.

In practice, generally, the invoices received by the users include the amounts corresponding to the sanitation and treatment of the water in addition to the amount for the supply service. Moreover, users are also charged for other concepts, such as fees which, as previously mentioned, in the majority of cases do not correspond to the municipality. This is the case, for example, of the charges for water treatment, transfers, local improvement, solid waste management, etc. (Babiano 2020, 19–20).

Each municipality determines the periodicity of the sending and collection of the bills, which is usually bi-monthly or quarterly. The variable fee is typically calculated through the periodical reading of the measurement devices installed, although, in some cases, estimated consumption can be used based on previous readings.

4.3 *Tariff Systems and the Challenge of the Multiplicity of Objectives*

The setting of the tariff structure of the urban water services is particularly important because, as indicated in the Water Framework Directive, it has to contribute to the achievement of economic, social and environmental objectives, which can sometimes be conflicting.[4] In fact, addressing all of these goals in a single tariff instrument is complex, giving rise to the emergence of a wide

4 See Suárez-Varela (2020) and Arbués and García-Valiñas (2020) for a review of studies analysing the effects of the different tariff designs on efficiency, sustainability, and equity.

range of price systems that seek to respond to the conflicts between these objectives (Suárez-Varela 2020, 207).

Fixed fee tariffs are the simplest. They do not require the installation of meters; they are easily administered and ensure stability in the revenue of the supplying entity. However, as they do not take into account the volume of consumption, this type of tariff does not encourage an efficient consumption of water and does not contemplate the payment capacity of the users, so it could be considered as being unequal. The monomial tariffs, meanwhile, take into account the real consumption of water, they have weaknesses in terms of efficiency and equity, given that there is a single price per cubic metre consumed and excessive consumption is not penalised.

The two-part or binomial systems are the most frequently used (García-Rubio, Ruiz-Villaverde, and González-Gómez 2015, 1461). In these cases, it is essential to determine a balance between the fixed and variable parts. In principle, the fixed component is included with the idea of contributing to cost recovery and the financial stability of the service providers. For this reason, it should be high enough to cover the fixed costs of the service which are usually higher than the variable costs. However, in practice, the weight of the variable fees tends to be reinforced in order to fulfil the criteria of efficiency and equity. If the fixed fee were excessively high, the tariff would have a regressive nature, given that the average cost per cubic metre consumed would be higher for low consumption than for high consumption, which would discourage responsible consumption (Hoque and Wichelns 2013, 478).

With respect to the variable part, the formula which has been long recommended is the increasing block system, given that it enables the equity principle to be followed while encouraging efficient consumption and environmental sustainability (Hoque and Wichelns 2013, 479; Instituto Aragonés del Agua 2013, 22; OECD 2009, 16). In this case, the underlying idea is that water satisfies different needs depending on the volume of consumption. In this way, a lower price per cubic metre is applied to the lowest level considered to be essential. Progressively higher prices are applied to consumption above this volume. This enables poorer households to access a minimum amount of water at a below-cost price, while households with a higher level of consumption are charged a higher price which compensates the previous deficit. In this way, crossed subsidising takes place between the households with the lowest and highest consumption (Boland and Whittington 2000, 5).

Increasing block tariffs also have their disadvantages. This system applies an additional charge to large families and those communities, usually the poorest, where several households share the same connection. In these cases, higher blocks are reached, with higher prices, although an abusive use of water is not actually made (OECD 2009, 91; Whittington 1992, 76). For this tariff system to be effective, it should fulfil at least two conditions. First, that the poorer households actually consume less water than higher-income households. This does not necessarily occur, given the low elasticity between the level of income and water consumption (Andres et al. 2019, 33; Nauges and Wittington 2017, 128; Boland and Wittington 2000, 13). Second, in order to ensure the cross subsidising between the two groups of consumers, the price of the higher blocks would have to be above the average cost, which would enable the tariff of the first block, that of basic consumption, to be reduced below this threshold. Similarly, in the intermediate blocks, the price should be the same as the marginal cost in order to encourage saving and favour cost recovery. However, in most cases, the price of all of the blocks, including those corresponding to the highest consumption, are below the average cost and the threshold of the first step, that of basic consumption, is usually much higher than the amount than that to satisfy minimum needs. This would mean that middle-income households would receive more subsidies than low-income households (Nauges and Wittington 2017, 134; Boland and Wittington 2000, 13). In order to avoid these disadvantages, different compensatory mechanisms have emerged. In some cases, minimum exempt amounts are applied, for which a certain amount of water, which usually coincides with the first block, is exempt from payment. In other cases, the tariffs are modified so as to take into account the size or level of income of the family unit. Also, connection subsidies are used for households with less resources, eliminating the fixed part of the tariff corresponding to the connection to the service. Programmes for transfers to households can also be used instead of applying redistribution through tariffs (Suárez-Varela 2020, 209–10).

All of the above illustrates the complexity of fulfilling the objectives of efficiency, equity, cost recovery and sustainability in a single tariff instrument, hence the existence of a wide variety of tariff structures and price systems.

5 Tariffs for the Supply of Water for Domestic Use in Andalusia

With almost 8.5 million inhabitants, Andalusia is the largest Autonomous Region in Spain in terms of population. It has a total of 785 municipalities distributed across eight provinces (INE 2022). Given the diversity in terms of types of management and tariff systems, analysing all of them is a truly extensive task. Therefore, in this section, the tariff systems of only the capitals of the eight Andalusian provinces are analysed.[5] Specifically, the different entities that manage the supply and sanitation services in these cities and the tariff systems used are examined.

5.1 *Regulatory Framework*

The urban water sector is highly regulated at all administrative levels (European, state, regional and local), giving rise to a complex regulatory framework and, in many cases, one which is not very transparent (CNMC 2020, 6).

From a European perspective, there is no water policy per se. The regulations are incorporated in the environmental policy of the European Union (EU). This competence is shared between the EU and the member states and is materialised in distinct environmental provisions and the principal objective is the conservation of the quality of the water and the sustainability of its management. Particularly noteworthy is the Water Framework Directive (WFD),[6] which establishes a community action framework within the field of water policy, and which constitutes a new water management model in the EU. Its main aim is to protect water in all its forms, for which it establishes a series of environmental objectives to be met within a term of 15 years. Another fundamental element of this Directive is the cost recovery principle in water services related to the idea that the polluter pays.

The WFD was transposed into Spanish legislation through Law 32/2003 of 30 December regarding fiscal, administrative, and social order measures which, in its article 129, includes the modification of the Restated Text of the Water Law. On a state level, the principal regulatory norm for water is the Royal Legislative Decree 1/2001 of 20 July, approving the Restated Text of the Water Law (TRLA).

In Spain, the different administrative levels intervene in water management in accordance with the competencies attributed to them by the Spanish Constitution, the different autonomy statutes, and Law 7/1985 of 2 April, regulating the local government rules (LRBRL). This law assigns the local entities with the competencies for the urban water cycle. However, in many cases, the regional regulations modify this concept.

In the case of Andalusia, Law 9/2010 of 30 July on Water of Andalusia, devotes a title (Title I) to the Administration of Water in Andalusia, in which

5 In the year 2020, an extensive study was published (Babiano 2020) on the urban water supply services in different parts of Spain and Europe, with a special focus on the Autonomous Region of Andalusia. The objective of this section is to present an updated overall view of the tariff systems in the Andalusian provincial capitals with a similar approach to the first part of the cited study.

6 Directive 2000/60/EC of the European Parliament and Council of 23 October.

the competencies of the regional government and local entities are specified. Among other tasks, the local governments are responsible for the downstream water supply services (distribution, intermediate storage and drinking water supply to the individual inlets or facilities for consumption) and sanitation (collection of urban wastewater and rainwater from the population nuclei through municipal sewage networks and their conveyance to the collection point for their treatment) (art. 16). Meanwhile, the domestic supply of water in Andalusia is subject to Decree 120/191 of 11 June, which approves the Regulations for the Domestic Supply of Water in this region. This provision regulates the relationships between the entity providing the domestic water supply service and the subscribers, indicating the basic rights and obligations of both parties.

On a local level, the regulation of the provision of these services, particularly the tariff systems, is carried out through municipal by-laws.[7] These are the principal sources used for studying the tariff structures in the Andalusian provincial capitals. The results of this analysis are presented below, grouped by concepts.

5.2 *Management Entities and Municipal By-Laws*

A total of seven companies provide the supply service in the provincial capitals of Andalusia:

- Almeria and Jaen: in these two cities, the service is managed by *Aqualia*, a company forming part of the private group FCC.
- Cadiz: the management corresponds to *Aguas de Cádiz* S.A., a municipal company, in the form of a public limited company whose capital is wholly state-owned.
- Cordoba: the urban water cycle is the responsibility of EMACSA (*Empresa Municipal de Aguas de Córdoba,* S.A.), a company that is 100% owned by the city council.
- Granada: the supply is administered by EMASAGRA (*Empresa Municipal de Aguas de Granada*), a mixed company with indirect management in which the City Council of Granada (51%) and *Hidralia, Gestión Integral de Aguas de Andalucía* S.A. (49%) participate.
- Huelva: the water supply services are provided by *Aguas de Huelva*, a municipal company in the form of a mixed company whose capital is owned by

7 The municipal by-laws consulted: Almeria Town Hall 2019, Cadiz Town Hall 2019, Cordoba Town Hall 2021, Granada Town Hall 2012, Huelva Town Hall 2021, Jaen Town Hall 2014, Malaga Town Hall 2022, and Seville Town Hall 2020.

the City Council of Huelva (51%) and *Hidralia, Gestión Integral de Aguas de Andalucía, S.A.* (49%).
- Malaga: the service is managed by EMASA (*Empresa Municipal de Aguas de Málaga*), a municipal company in the form of a limited company, 100% owned by the City Council of Malaga.
- Seville: the management of the urban water cycle corresponds to EMASESA (*Empresa de Abastecimiento y Saneamiento de Aguas de Sevilla S.A.*), a metropolitan company owned by the eleven local governments to which it provides services, with the majority shareholder being the City Council of Seville (69% of the capital).

The conditions for providing these services and, particularly, the tariffs to apply are set forth in the corresponding by-laws. In many cases, in the years following the approval of the by-law, updates of the tariffs are published in order to adjust to the overall increase in prices.

The design of the tariffs and the proposals for their modification correspond to the City Council, which should approve them in a Plenary Session and record them in a by-law. However, given that the competencies in terms of price intervention correspond to the Regional Government, when the local entity decides to update the tariffs, it needs the approval of the Regional Government of Andalusia (Ordóñez 2002, 96).

In all of the Andalusian provincial capitals, the supply services are provided through companies and not directly by the local government. Therefore, the tariffs charged take the form of the provision of a public untaxed asset, which, as such, should be approved by the regional government. This is stated in all of the by-laws consulted, except for those of Jaen and Granada, where the latest regulation of this economic figure was implemented before 2017.[8]

Table 10.1 shows the most relevant information regarding the type of management, the responsible entities and the dates of approval of the municipal by-laws regulating the tariffs of the drinking water supply services in the eight Andalusian provincial capitals. An analysis of Table 10.1 reveals that, although in none of the cities studied the supply is directly provided by the local entity, there is wide diversity with respect to the type of managing entities of the service, with public, mixed, and private companies.

5.3 *Tariff Systems*

An analysis of the by-laws relating to the water supply service (for domestic use) reveals that in the eight cities studied the tariff structure is binomial,

8 Provision of public untaxed assets, a concept introduced in 2017 through Law 9/2017, of 8 November, on Public Procurement Contracts.

TABLE 10.1 Water supply service: managing entities, by-laws and updates of tariffs

City	Managing entity	Type of management	Date of municipal by-law[a]	Date of last tariff update[b]
Almeria	*FCC Aqualia*	Indirect (Private)	April 2019	December 2021
Cadiz	*Aguas de Cádiz, S.A.U.*	Direct (Public)	October 2019	June 2020
Cordoba	*EMACSA*	Indirect (Mixed)	September 2020	April 2021
Granada	*EMASAGRA*	Indirect (Mixed)	December 2012	March 2014
Huelva	*Aguas de Huelva*	Indirect (Mixed)	September 2021	February 2022
Jaen	*FCC Aqualia*	Indirect (Private)	December 2010	August 2014
Malaga	*EMASA*	Direct (Public)	February 2022	February 2022
Seville	*EMASESA*	Direct (Public)	May 2015	May 2015

a Municipal by-law regulating the tariffs for the provision of the water supply service.
b Water supply service tariffs.
SOURCE: OWN ELABORATION WITH DATA FROM BABIANO (2020, 68–88) AND THE MUNICIPAL BY-LAWS REGULATING THE TARIFFS FOR THE PROVISION OF THE WATER SUPPLY SERVICE

that is, with two components: a fixed part for the provision of the service and another variable part depending on consumption, organised in increasing blocks.

The model most used is that of bi-monthly billing with a fixed component modulated in accordance with the calibre of the meter installed (usually measured in millimetres), and another variable part organised into four increasing blocks, delimited in accordance with the cubic metres of water consumed, although with different limits in each city. However, the way each of them is determined is not the same in all of the cities studied. Table 10.2 shows the principal characteristics of the tariff structures of the water supply services for the eight Andalusian provincial capitals. Observing the information contained in the table, it should be noted that although the most used parameter for determining the fixed amount is the calibre of the meter, there are two cities (Cordoba and Jaen) that also take into account the location of the building on the fiscal map and one (Almeria) that establishes a fixed amount irrespective of the meter.

By way of example, Figure 10.1 shows the monthly cost of the fixed amount in the eight provincial capitals of Andalusia taking a 20 mm meter as a

TABLE 10.2 Tariff structure of the supply of water for domestic use in the Andalusian provincial capitals

City	Fixed component (service)	Variable component (consumption)		Frequency of invoicing
	Parameter used	No. of blocks	Parameter for determining the blocks	
Almeria	Fixed amount	4	m^3 of water consumed	Quarterly
Cadiz	Calibre of meter	4	m^3 of water consumed	Bi-monthly
Cordoba	Calibre of meter and fiscal situation of the supply point	3	m^3 of water consumed	Bi-monthly
Granada	Calibre of meter	4	m^3 of water consumed	Bi-monthly
Huelva	Calibre of meter	4	m^3 of water consumed	Bi-monthly
Jaen	Calibre of meter and category of the street of the building (5 categories)	4	m^3 of water consumed (different tariffs for each street category)	Quarterly
Malaga	Calibre of meter	4	m^3 of water consumed and number of household members	Bi-monthly
Seville	Calibre of meter or number of households (multiple meter)	3	m^3 of water consumed per inhabitant and month	Quarterly

SOURCE: OWN ELABORATION WITH DATA FROM THE MUNICIPAL BY-LAWS REGULATING THE TARIFFS FOR THE PROVISION OF THE WATER SUPPLY SERVICE

reference. For the case of Cordoba and Jaen, which take into account the location of the household, an intermediate category has been considered, level 4 of 7 for Cordoba and 3 of 5 for Jaen. The wide disparity in prices is striking, with Cordoba having prices much higher than the average and Granada with the lowest value.

On the other hand, with respect to the variable amount, the blocks are determined in all cases according to consumption measured in cubic metres. However, two cities (Malaga and Seville) also take into account the number of household members as a more precise way to calculate the blocks for each subscriber. The

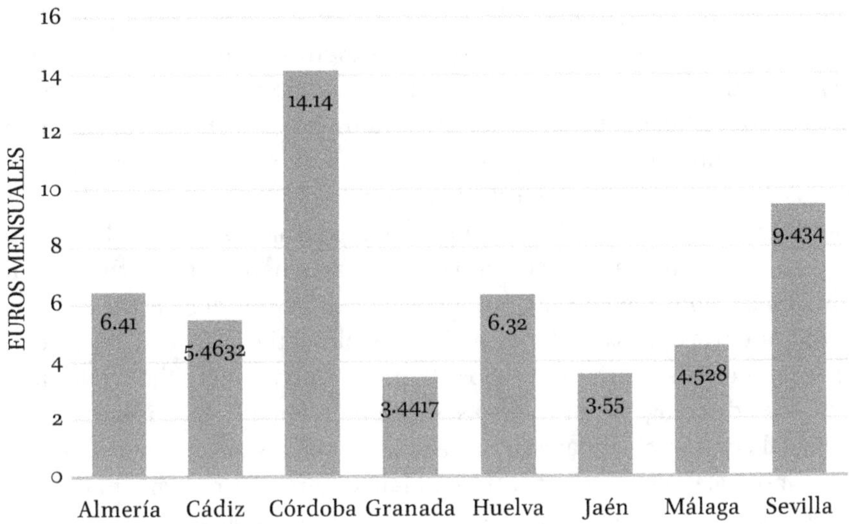

FIGURE 10.1 Price of the fixed amount for domestic water supply services (20 mm calibre meter) (Euros per month)
SOURCE: OWN ELABORATION WITH DATA FROM THE MUNICIPAL BY-LAWS REGULATING THE TARIFFS FOR THE PROVISION OF THE WATER SUPPLY SERVICE

Annex to this chapter includes a breakdown of the tariffs by consumption block for the different cities. Conducting a comparative study between them would be complex given the diversity of the range and prices and the difficulty of taking into account all of the possible variants listed in the by-laws. Including such a comparison in this study would make it excessively long. However, the study by Babiano (2020) makes a comparative analysis of these prices with the tariffs in force at the time of its elaboration and under the assumption that many parameters are needed to make this comparison possible.

5.4 *Bonuses*

In the majority of the cities, the managing entities apply bonuses to the tariffs of certain social groups. The most frequent are those applied to pensioners, large families, or those at risk of social exclusion and the unemployed. However, sometimes the by-laws also include reductions and special treatment for other groups or through special mechanisms. A summary of the bonuses applied to the tariffs of the supply of water for domestic use in the cities analysed can be seen in Table 10.3. Three cities (Cadiz, Cordoba, and Huelva) apply a Minimum Subsistence Supply for families at risk of social exclusion, which guarantees that nobody is deprived of water due to economic reasons. It works in a similar way in all cases. Those families who can accredit their situation of difficulty can access a minimum volume of consumption at no cost: 6 cubic metres per person and bi-monthly invoice in Cadiz; 9 cubic metres per person and for three months in Cordoba and 3 cubic metres per person and month in Huelva. On the other hand, two cities, Cordoba, and Seville, apply bonuses for efficient use, reducing the tariffs for those households with low consumption (less than 6 cubic metres per household and quarter in Cordoba and 3 cubic metres per inhabitants and month in Seville). In the case of Malaga, the managing entity (EMASA) has a Social Fund with economic resources to help those clients suffering from economic problems and who have a debt with the company. Finally, from Table 10.3, it may be deduced that Cordoba is the city with the widest range of possible bonuses. Meanwhile, at the other end of the scale, Jaen only provides assistance to those households with more than four members in a wider range of tariff blocks. It can therefore be established that the provincial capitals of Andalusia have similar tariff systems as they all implement binomial structures, but with sufficient differentiating elements so as to make a comparative analysis between them highly complex.

TABLE 10.3 Bonuses to the tariffs of the supply of water for domestic use

	Beneficiaries of the bonuses				
Cities	**Pensioners**	**Large families**	**Families at risk of social exclusion**	**Unemployed**	**Other**
Almeria	x	x	x		
Cadiz					– Minimum subsistence supply
Cordoba	x	x	x	x	– Bonus for low domestic consumption (general) – Minimum subsistence supply
Granada	x	x		x	
Huelva	x	x		x	– Bonuses for the disabled and victims of gender-based violence – Minimum subsistence supply
Jaen					– Bonuses for households with more than 4 members
Malaga					– EMASA social fund
Seville			x		– Bonus for efficient use (general)

SOURCE: OWN ELABORATION WITH DATA FROM THE MUNICIPAL BY-LAWS REGULATING THE TARIFFS FOR THE PROVISION OF THE WATER SUPPLY SERVICE

6 Conclusions

Currently, as a consequence of water scarcity and the need to guarantee its access to the entire population, the management of water resources has to fulfil multiple objectives, which sometimes can seem to be in conflict with one another. From the point of view of establishing water prices, it is not easy for a single tariff instrument to combine economic efficiency, environmental sustainability and social equity. For this reason, many types of structure have emerged seeking to respond to the conflicts arising between these objectives.

The most frequent are two-part tariff systems, or binomial tariffs, with an increasing block structure. This model includes, on the one hand, a fixed component with the idea of contributing to cost recovery. On the other hand, it also includes a variable element based on the establishment of increasing blocks, allowing the principal of equity to be obtained while encouraging efficient consumption and environmental sustainability.

All of the Andalusian provincial capitals apply this binomial model with increasing blocks in their tariffs for the supply of water for domestic use. However, within this basic structure, each of them has many variants. They all use the calibre of the meter to determine the fixed amount and the consumption of water to determine the variable part, complemented, in some cases, with the location of the building or the number of members in the household. However, there are too many differential elements in terms of sections to determine the fixed component, number and thresholds of the blocks, prices for the sections and bonuses applied so as to carry out a comparative analysis between them with definitive conclusions. Such an analysis with the necessary academic rigour would be too extensive to include in this study.

Municipal By-Laws

Almeria Town Hall. 2019. *Ordenanza reguladora de la prestación patrimonial de carácter público no tributario por suministro domiciliario de agua potable.* BOP de Almería, no. 200, 17 May 2019, 2–5. Last tariff update: BOP Almería no. 42, 3 March 2022.

Cadiz Town Hall. 2019. *Ordenanza no. 41 reguladora de las tarifas por servicios de abastecimiento, saneamiento y depuración de aguas de la ciudad de Cádiz.* BOP de Cádiz no. 209, 31 October 2019. Last tariff update: BOJA no. 176, 10 September 2020.

Cordoba Town Hall. 2021. *Ordenanza no tributaria reguladora de la prestación patrimonial de carácter público no tributaria a percibir por EMACSA por los servicios de abastecimiento de agua potable y otras actividades relacionadas con los mismos, así como de determinados aspectos técnicos de las instalaciones y de ciertos riegos y suministros específicos.* BOP Cordoba no. 72, 19 April 2021.

Granada Town Hall. 2012. *Ordenanza reguladora de la tarifa por prestación del servicio de abastecimiento domiciliario de agua potable y otras actividades conexas al mismo.* BOP Granada no. 246, 24 December 2012. Last tariff update: BOP Granada BOJA no. 54, 20 March 2014.

Huelva Town Hall. 2021. *Ordenanza municipal reguladora de la prestación patrimonial de carácter público no tributario por prestación del servicio de abastecimiento de agua y otros derechos económicos por actividades conexas al mismo en el municipio de Huelva.* BOP Huelva no. 183, 24 September 2021. Last tariff update: BOJA no. 34, 18 February 2022.

Jaen Town Hall. 2014. *Ordenanza fiscal reguladora de la tasa por la prestación de los servicios vinculados al ciclo integral del agua.* BOP Jaén no. 291, 21 December 2010. Last tariff update: BOP Jaén no. 160, 21 August 2014.

Malaga Town Hall. 2022. *Ordenanza reguladora de la prestación patrimonial de carácter público no tributario por la prestación del servicio de abastecimiento de agua potable, y otros servicios conexos.* BOP Málaga no. 23, 3 February 2022.

Seville Town Hall. 2020. *Ordenanza reguladora de la prestación patrimonial de carácter público no tributario del servicio, prestado por Emasesa, de Abastecimiento domiciliario de agua potable, Saneamiento (Vertido y Depuración) y otras actividades conexas a los mismos.* BOP Sevilla no. 104/2020, 7 May 2020.

Annex: Variable Charges of Domestic Water Consumption in the Provincial Capitals of Andalusia

(In accordance with the by-laws regulating the service and updating the tariffs)

1. Tariff systems of quarterly consumption

	Almeria		Jaen	
	Quarterly consumption	Euros/m³	Quarterly consumption	Euros/m³
Block I	0–15 m³	0.524281	0–18 m³	0.292
Block II	15–50 m³	0.873797	19–36 m³	0.552
Block III	50–90 m³	1.478731	37–54 m³	1.108
Block IV	Over 90 m³	3.535507	Over 54 m³	2.041

2. Tariff systems of bimonthly consumption

	Cordoba	
	Bimonthly consumption	Euros/m³
Block I	0–18 m³	0.793
Block II	18–36 m³	1.0309
Block III	Over 36 m³	1.2292

3. Tariff systems of monthy consumption

	Cadiz		Granada		Huelva	
	Monthly consumption	Euros/m³	Monthly consumption	Euros/m³	Monthly consumption	Euros/m³
Block I	0–3 m³	0.4259	0–2 m³	0.4053	0–2 m³	0.520
Block II	3–7 m³	0.6343	2–10 m³	0.6763	2–9 m³	0.835
Block III	7–11 m³	0.7181	10–18 m³	1.3996	9–15 m³	1.355
Block IV	Over 11 m³	0.2786	Over 18 m³	1.9171	Over 15 m³	1.901

4. Tariff systems of monthly consumption per inhabitant

	Malaga		Seville	
	Monthly consumption / Inhabitant	Euros/m³	Monthly consumption /inhabitant	Euros/m³
Block I	0–2 m³	0.191	0–4 m³	0.542
Block II	2–3 m³	0.492	5 m³	0.920
Block III	3–5 m³	0.713	Over 5 m³	1.749
Block IV	Over 5 m³	1.406		

References

AEAS and AGA. 2020. XVI *Estudio Nacional de Suministro de agua potable y saneamiento. Datos del sector del agua urbana 2020*. Accessed 22 March 2022. https://www.aeas.es/images/Doc_Es_Nacional/2020/2021-03-31_-_Trptico_completo_para_web_con_logos_-_publicado.pdf.

Álvarez García, S., M. A. García Valiñas, and J. Suárez Pandiello. 2001. "Tarifas no uniformes: Servicio de suministro doméstico de agua." *Documentos del Instituto de Estudios Fiscales*. DOC (8/03).

Andres, L. A., M. Thibert, C. Lombana Cordoba, A. V. Danilenko, G. Joseph, and C. Borja-Vega. 2019. *Doing More With Less: Smarter Subsidies for Water Supply and Sanitation*. Washington, DC.: World Bank. https://doi.org/10.1596/32277.

Arbués, F. and M. García-Valiñas. 2020. "Water tariffs in Spain". *Oxford Research Encyclopedias*. https://doi.org/10.1093/acrefore/9780190632366.013.246.

Babiano, L. (dir.). 2020. *Tarificación y benchmarking. Estudio sobre los servicios urbanos de abastecimiento y saneamiento de aguas*. AEOPAS and EMASESA.

Boland, J. J. and D. Whittington. 2000. "Water tariff design in developing countries: disadvantages of increasing block tariffs (IBTs) and advantages of uniform price with rebate (UPR) designs." *World Bank Water and Sanitation Program, Washington, DC*. 37.

CNMC. 2020. *Estudio sobre los servicios de abastecimiento y saneamiento de agua urbana*. Colección Estudios de mercado E/CNMC/07/2019.

Daza Sánchez, F. 2008. "Demanda de agua en zonas urbanas de Andalucía." PhD diss., University of Cordoba.

García-Rubio, M. A., A. Ruiz-Villaverde, and F. González-Gómez. 2015. "Urban water tariffs in Spain: What needs to be done?" *Water* 7 (4): 1456–79. https://doi.org/10.3390/w7041456.

González Gómez, F. G. 2005. "El precio del agua en las ciudades. Reflexiones y recomendaciones a partir de la Directiva 2000/60/CE." *Ciudad y territorio: estudios territoriales* 144: 305–320.

Hoque, S. F. and D. Wichelns. 2013. "State-of-the-art review: designing urban water tariffs to recover costs and promote wise use." *International Journal of Water Resources Development* 29 (3): 472–91. https://doi.org/10.1080/07900627.2013.828255.

INE. 2018. *Distribución de agua registrada, usuario y periodo, año 2018*.

INE. 2022. *Número de municipios por comunidad autónoma y provincia*.

Instituto Aragonés del Agua. 2013. *Tarifas de agua*. Zaragoza: Instituto Aragonés del Agua.

Nauges, C. and D. Whittington. 2017. "Evaluating the performance of alternative municipal water tariff designs: Quantifying the tradeoffs between equity, economic efficiency, and cost recovery." *World Development* 91: 125–43. https://doi.org/10.1016/j.worlddev.2016.10.014.

OECD. 2009. *Managing water for all. An OECD perspective on Pricing and Financing.* Paris: OECD.

Ordóñez de Haro, C. 2002. "Análisis de las tarifas de abastecimiento doméstico de agua en la provincia de Málaga: una aproximación." *Revista de Estudios Regionales* (64): 93–121.

PWC. 2018. *La gestión del agua en España. Análisis y retos del ciclo urbano del agua.* Accessed 12 April 2022. https://www.pwc.es/es/publicaciones/energia/assets/gestion-agua-2018-espana.pdf.

Suárez-Varela, M. 2020. "Las políticas de tarifas urbanas del agua en España." *Presupuesto y Gasto Público* 101: 205–26.

Whittington, D. 1992. "Possible adverse effects of increasing block water tariffs in developing countries." *Economic Development and Cultural Change* 41 (1): 75–87. https://doi.org/10.1086/451996.

CHAPTER 11

Research on Wastewater Taxes: An Essential Contribution to Improving Environmental Management

Leticia Gallego-Valero, Encarnación Moral-Pajares and Isabel María Román-Sánchez

Abstract

Taxes on wastewater are an essential tool for the proper management of water resources, contributing to the planet sustainability. The bibliometric analysis carried out using WoS shows the need for more scientific contributions on wastewater taxes, given the insufficient results in this research field in the period 2000–2020. A worldwide search yielded 144 results, representing less than 1.5% of the total research on all environmental taxes. Spain, the USA, and China account for 50.68% of the total. This paper is aimed at researchers in this area and seeks to help them understand the current research environment and identify future trends.

Keywords

bibliometric analysis – environmental management – water resources – wastewater taxes – WoS database

1 Introduction

Given the scarcity of and increased demand for water resources, there is a need for better water management and more sustainable production processes (Vázquez et al. 2015; Burak and Margat 2016). The importance of water and wastewater for human development, the environment and the economy has led the UN to declare access to water and sanitation a basic human right, while the OECD's Environmental Outlook for 2050 addresses the need to protect water resources. The ever-increasing demand for water is driven by factors such as population growth in urban areas, economic development, industrialisation, energy production, improvements in health and sanitation systems,

 | DOI:10.1163/9789004679634_012

rising agricultural productivity and the expansion of irrigation systems in arid regions (Uche et al. 2015; Vörösmarty et al. 2000). Indeed, global water use has increased sixfold over the past 100 years and continues to rise at a steady rate of 1% annually (UN-Water 2020). This has exacerbated water scarcity and water stress, prompting a search for new ways to manage this resource. Recent innovations in this regard include the Nasdaq Veles California Water Index (NQH2O), the first index for water prices in California, along with the creation of a futures market, or the reuse of wastewater as a reliable alternative source (Connor et al. 2017).

By improving wastewater management, the aim is not only to reduce the pollution of water sources, but also to remove pollutants from wastewater flows, make use of reclaimed water and recover useful by-products (Weerasekara 2017). Doing so, however, requires the implementation of fees and taxes; for example, wastewater taxes are economic or incentive-based instruments used as environmental policy tools (Bluffstone 2003). These instruments seek to alter behaviour through the use of (dis)incentives rather than directives, in the hope that environmental outcomes will be achieved more efficiently. The management of these instruments varies by country, and they may be levied at the national, regional, or municipal level (Zárate, Vallés, and Trueba 2007). Environmental taxes are paid without the promise of receiving anything in return; they are intended as a way to correct a negative externality, regardless of whether the revenues collected are directed towards improvements in the natural environment (Riera 2005). Generally speaking, there are three possible reasons for levying a tax on water: to reduce the consumption of this natural resource and, at the same time, discharges; to influence the behaviour of polluting actors; and to contribute to financing the investment and operating costs of the infrastructure required for the treatment (Gallego-Valero, Moral-Pajares and Román-Sánchez 2018).

The present study is a bibliometric analysis of taxes levied on wastewater. Bibliometrics, which is the use of mathematical and statistical methods to analyse scientific publications, can be applied to all fields of knowledge and all types of publications (Pérez et al. 2017). The use of the internet for data collection is accepted by the scientific community (Bryman and Bell 2015) and there are several databases that are commonly used for the bibliometric evaluation of research, notable among which is the Web off Science (WoS) platform (Archambault, Campbell, and Gingras 2009).

Specifically, this study is a quantitative and qualitative analysis of the dynamics of global research on wastewater taxes in the period 2000–2020. It explores the existing research on this subject, the most relevant areas of research in this field and the limits in the related research. To that end, a review was carried out of the existing literature on wastewater taxes published in the period under study, applying a systematic qualitative analysis and a quantitative bibliometric analysis to identify, organise and study the most relevant characteristics of the research, through the use of the WoS database and statistical processing tools. Thus, this paper collects and compiles all the information regarding research on this subject in the last 21 years, the ultimate aim of which is to show the evolution of the research, the main countries and institutions that work on the subject analysed, and the leading journals publishing this research. The results can be of use for the scientific community by offering a better understanding the current research environment and identifying future trends that can help guide researchers when undertaking a new study.

2 Materials and Methods

2.1 *Bibliometric Analysis*

Bibliometric analysis is a research method that makes it possible to identify, organise and evaluate the constituent elements of a specific area of study, and is thus an important tool for literature reviews (Huang et al. 2014; Aznar-Sánchez et al. 2018). It involves various different types of indicators relating to the quantity (productivity), quality (impact of publications) and structure (analysing connections) of research (Durieux and Gevenois 2010). The bibliometric analysis performed in this article uses the WoS database from Clarivate Analytics. WoShas high visibility in the different areas of knowledge, a selection filter for prestigious publications, and is also widely used to carry out bibliometric studies (Prigle 2008). Previous studies have used bibliometric analysis to explore areas such as the use of water or wastewater (Montoya et al. 2016; López-Serrano et al. 2020) or renewable energies, sustainability, and the environment (Romo-Fernández, Guerrero-Bote, and Moya-Anegón 2013).

2.2 *Data Selection and Processing*

The sample of documents analysed in this study was obtained by conducting a search of the entire WOS database with the term "wastewater taxes" in the option "topic":

- The results of the search were compared with those for other environmental taxes, such as taxes on hydrocarbons, electricity, motor vehicles, gas emissions and waste.
- The period analysed was 2000–2020. The analysis yielded a final sample of 144 results on wastewater taxes. The sample selection was done in January 2021.
- The following variables were analysed: evolution, areas of study, main countries, main journals, main institutions, and main authors.

Bibliometric studies distinguish between three types of indicators (Durieux and Gevenois 2010): quantity indicators, which refer to productivity; quality indicators, which refer to the impact of publications; and structural indicators, which measure the connections established between the different agents (authors, institutions, and countries). In this study, quantity and quality indicators were analysed, but not structural indicators. In addition to the measure of productivity of the countries, institutions and authors, the following indicators were used to evaluate the quality of the journals in which the documents were published:

- SJR (SCImago Journal Rank): measure of the scientific influence of scholarly journals that accounts for both the number of citations received by a journal and the importance or prestige of the journals where the citations where made (Trinh et al. 2013).
- Quartile in which the journal is positioned.
- Number of citations.
- JIF (Journal Impact Factor): measure of the frequency with which the average article in a journal has been cited in a particular year.
- Total publications.
- h-Index: an author-level metric that measures both the citation impact of the publications and the productivity of an author (Falagas et al. 2008).

After selecting the sample, the information available in the WOS database was downloaded and prepared for analysis by eliminating duplications, correcting mistakes, and adding missing information. The first step in the analysis was to quantify the global and Spanish contribution to the research on different types of environmental taxes, and analyse the evolution of the field over the period 2000-2020. The main areas of study relating to wastewater research were identified by referring to WOS classifications, before establishing the leading contributing countries to this body of research. In the next step, the main journals were identified and listed with their SJR, JIF and total citations index for the year 2019 (to evaluate the impact of the journals). Lastly, the main institutions and most prolific authors were tabulated, along with their total number of publications, total citations, and h-Index per author as an indicator of productivity.

3 Results

3.1 *Global and Spanish Contribution to Research on the Different Types of Environmental Taxes*

Within the publications on environmental taxes, those on wastewater taxes represent the smallest share of the total. While the general search term "environmental taxes" yields 10,333 results, there are far fewer results for specific types of tax, with large differences depending on the nature of the environmental tax studied. Figure 11.1 shows the research on different environmental taxes by areas, showing two differentiated groups. The first group comprises the papers that account for the largest share of the search results for "environmental taxes", with these taxes having attracted the most attention from the scientific community – namely, electricity taxes (20.71%), gas emission taxes (20.56%) and waste taxes (15.41%), together totalling 56.68%. In the second group are those accounting for a smaller share – namely, motor vehicle taxes (6.46%), hydrocarbon taxes (3.44%) and wastewater taxes (1.39%). This clearly points to a niche opportunity for research, given the small number of scientific publications on these taxes.

Regarding the Spanish contribution to the research (Figure 11.2), the highest figure corresponds to wastewater taxes, where Spanish publications represent 11.11% of world research on wastewater taxes (16 articles of the 144 globally). Conversely, Spanish institutions have contributed relatively very little to the research on other environmental taxes, accounting for less than 5% in the other research areas identified (4.51% in hydrocarbon taxes, 3.60% in electricity taxes, 3.58% in waste taxes, 2.97% in gas emission taxes and 2.54% in motor vehicle taxes).

3.2 *Evolution of the Research on Wastewater Taxes*

Over the decade studied, the scientific community has shown very little interest in analysing taxes levied on wastewater, although the number of theoretical and applied researcher papers on the subject has gradually increased in that time. Figure 11.3 presents the evolution of articles published in this field from 2000 to 2020, as revealed by the bibliometric analysis. The search of the WOS database for "wastewater taxes" returns only 144 results – compared to the 10,333 articles on the topic of "environmental taxes"– clearly showing that this field of research remains underexplored. Indeed, in percentage terms, research on wastewater taxes comprises only 1.39% of the total for environmental taxes. That said, the research on wastewater taxes shows an increasing trend between 2000 and 2020, rising sharply from 2010, albeit with notable fluctuations. It can

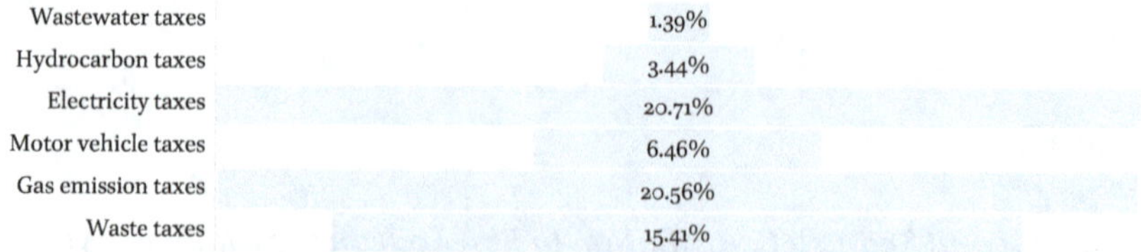

FIGURE 11.1 Environmental taxes by area as a percentage of total results for environmental taxes
SOURCE: OWN ELABORATION BASED ON WOS (2020)

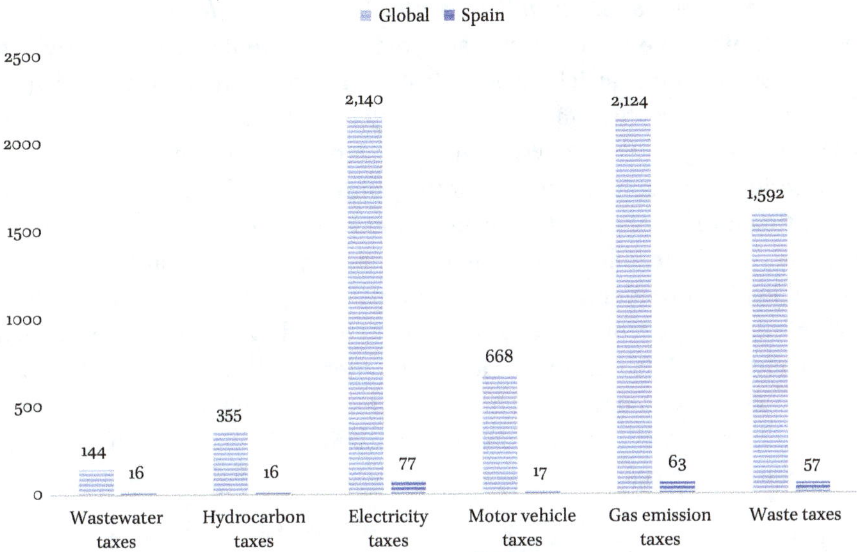

FIGURE 11.2 Global and Spanish scientific contribution to research on environmental taxes by area (2000–2020)
SOURCE: OWN ELABORATION BASED ON WOS (2020)

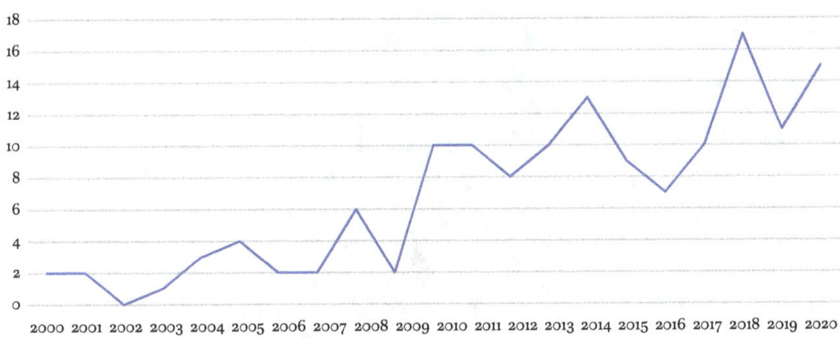

FIGURE 11.3 Trends in the research on wastewater taxes
SOURCE: OWN ELABORATION BASED ON WOS (2020)

be seen that the majority of articles are concentrated in the second half of the period studied, from 2010 on (120 studies, 83.3% of the total). A total of 2 articles were published in 2000 and 15 in 2020. The average year-on-year growth in the research on wastewater taxes is far higher than that for environmental taxes: the former registers an average year-on-year growth of 30.95%, while the corresponding figure for the latter is 12.97%. Therefore, the first finding of this paper is that research on wastewater taxes represents a very small proportion of the research on environmental taxes, although it has been increasing in importance, especially since 2010.

3.3 *Main Areas of Study in Wastewater Taxes Research*

The results of the bibliometric analysis reveal the different disciplines to which the analysed scientific articles belong. It should be noted that an article can be categorised in more than one discipline, which is why the results are analysed in percentages. Figure 11.4 shows the main areas of study in wastewater research in the period under study: predominantly Environmental Sciences and Ecology, accounting for 13% of the total, Business and Economics (10%) and Engineering (10%). These are followed by Water Resources (7%), Energy and Fuels (6%) and Public, Environmental and Occupational health (6%). The item "other" (48% of the total) encompasses 72 disciplines of widely varying subjects, such as Agriculture, Public Administration, Computer Science, Science Technology Other Topics, Toxicology, Biotechnology Applied Microbiology, Government Law, Biodiversity Conservation, Mathematics, Material Science, Meteorology Atmospheric Sciences.

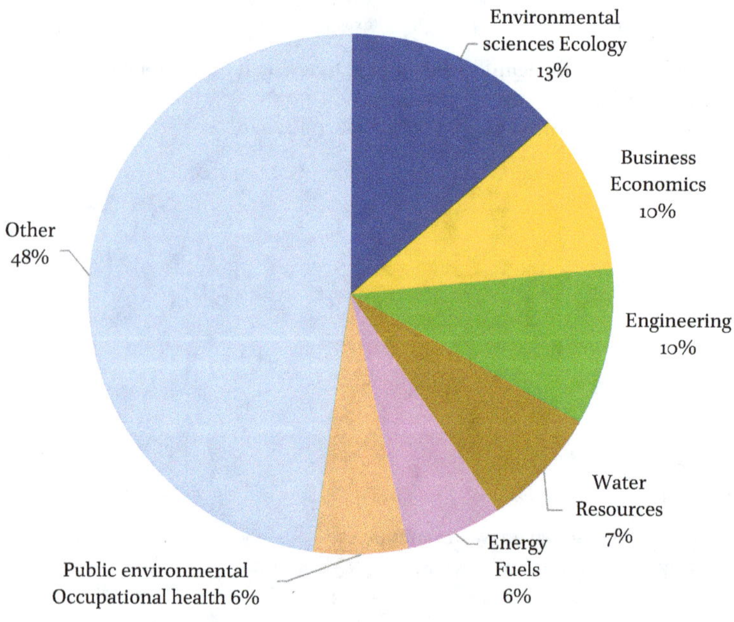

FIGURE 11.4 Main areas of study in wastewater taxes research (percentage)
SOURCE: OWN ELABORATION BASED ON WOS (2020)

3.4 *Relevant Countries in Wastewater Taxes Research*

The articles published on wastewater in the period 2000–2020 come from a total of 47 countries. In order of quantity of results, the top countries are the USA, China, Spain, Italy, Australia, Brazil, Denmark, Germany, Indias, Iran, Ireland, the Netherlands, Norway, Switzerland, Taiwan, Argentina, Austria,

Canada, Chile, the UK, Greece, Israel, Mexico, and Portugal. There are countries from almost all continents, with very diverse economic and sociocultural characteristics.

Figure 11.5 presents the top countries in the field. The USA is the most prolific country, representing 25% of the total, with 36 results. The second place is held by China (14.58% of the total, 26 articles). Third is Spain, with 16 results and 11.11%. These countries are followed by Italy (8 results, 5.56%), Australia, Brazil, Denmark, Germany, India, and Iran (4 results per country, 2.78% each) and Ireland, the Netherlands, Norway, Switzerland, and Taiwan (3 articles per country, 2.08% each).

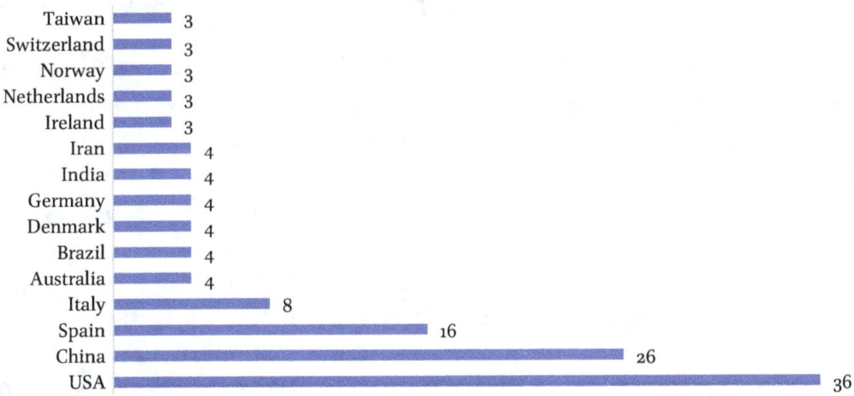

FIGURE 11.5 Top countries in wastewater taxes research
SOURCE: OWN ELABORATION BASED ON WOS (2020)

3.5 *Journals in Wastewater Taxes Research*

This section presents the most relevant journals publishing articles on wastewater taxes research and the corresponding indexes. Table 11.1 shows the main journals in terms of number articles, with the top positions held by Water Science and Technology (6 articles), Science of the Total Environment (4) and Desalination (4). Together, these journals published 9.72% of the total papers on this research subject. The journals listed are published in various different countries, with the United Kingdom, the Netherlands, Switzerland, and the USA being the most prolific.

It is important to note the broad range of journals publishing articles on wastewater taxes. The journal with the largest number of articles, Water Science and Technology, accounts for 4.16% of the total sample, and in 2019 had an SJR of 0.47 (Q2 quartile), a JIF of 1.638 and a total of 20,937 citations. The second is Science of the Total Environment, with 2.78% of the total sample, and in 2019 an SJR of 1.66 (Q1 quartile), a JIF of 6.551 and 134,962 total citations.

TABLE 11.1 Journals and relevant indexes in wastewater taxes research

Journal	Articles	SJR (2019)	Country	JIF (2019)	Total citations (2019)
Water Science and Technology	6	0.47 (Q2)	United Kingdom	1.638	20,937
Science of the Total Environment	4	1.66 (Q1)	Netherlands	6.551	134,962
Desalination	4	1.81 (Q1)	Netherlands	7.098	44,845
Waste Management	3	1.63 (Q1)	United Kingdom	5.448	30,533
Journal of Cleaner Production	3	1.89 (Q1)	Netherlands	7.246	104,138
Resources Conservation and Recycling	3	2.22 (Q1)	Netherlands	8.086	17,121
Water	3	0.66 (Q1)	Switzerland	2.544	13,460
Agricultural Water Management	2	1.37 (Q1)	Netherlands	4.021	17,178
Applied Energy	2	3.61 (Q1)	United Kingdom	8.848	98,570
Bioresource Technology	2	2.43 (Q1)	Netherlands	7.539	131,781
Desalination and Water Treatment	2	0.33 (Q2)	Italy	0.854	14,535
Energies	2	0.64 (Q2)	Switzerland	2.702	29,605
Environment International	2	2.36 (Q1)	USA	7.577	29,467
Environmental Engineering and Management Journal	2	0.32 (Q3)	Romania	1.186	2,223
Industrial & Engineering Chemistry Research	2	0.9 (Q1)	USA	3.573	79,542
International Journal of Life Cycle Assessment	2	1.6 (Q1)	Germany	4.307	8,278
Journal of Environmental Management	2	1.32 (Q1)	USA	5.647	44,264
Journal of the South African Institution of Civil Engineering	2	0.19 (Q4)	South Africa	0.540	167
Revista Publicando	2	-	Ecuador	-	-
Sustainability	2	0.58 (Q2)	Switzerland	2.576	35,095

SOURCE: OWN ELABORATION BASED ON WOS (2020)

In third place is Desalination, with 2.78% of the total sample and an SJR (2019) of 1.81 (Q1), a JIF index of 7.098 and 44,845 total citations. Below these three results there is a wide variety of journals, almost all of which present high index scores – most are in the first and second quartiles – and which together published 31.94% of the articles.

3.6 *Institutions and Authors in Wastewater Taxes Research*

Table 11.2 shows the top 19 authors in wastewater research by number of results, who are linked to 16 different institutions located in 10 different countries. The wide range of authors with only a few results per researcher indicates that this field of research is not very concentrated. Román-Sánchez I. has published the most articles (5), followed by Moral Pajares E. (3), Sánchez-Pérez J. A. (3) and Gallego-Valero L. (3), with co-authorship links in some of the results. There are some high-level scientists working in this field of study, as indicated by their total number of publications, total citations, and their h-Index. As can be seen, there

TABLE 11.2 Main authors in wastewater taxes research

Authors	Articles	Country	Affiliation	Total publications	Total citations	h-Index
Román-Sánchez, I.M.	5	Spain	University of Almeria	18	200	6
Moral-Pajares, E.	3	Spain	University of Jaen	25	57	4
Sánchez-Pérez, J.A.	3	Spain	University of Almeria	114	4,638	32
Gallego-Valero, L.	3	Spain	University of Jaen	6	9	2
Molinos-Senante, M.	2	Chile	Pontificia Univ Catolica Chile	113	2,493	25
Araujo, J.	2	Portugal	Centro Valorizacao Residuos, Guimaraes	12	83	3
Aydiner, C.	2	Turkey	Gebze Technical University	32	1,228	16
Birol, E.	2	USA	Int Food Policy Res Inst.	49	1,128	14
Carvalho, J.	2	Portugal	Minho University	14	289	10
Castro, F.	2	Portugal	Minho University	49	736	16
Das, S.	2	India	Madras Sch Econ	9	105	4
Eksioglu, S.D.	2	USA	Mississippi State Univ	48	1,261	18
El-Halwagi, M.M.	2	USA	Texas A&M Univ	487	10,192	51
Farrizi, M.	2	Iran	Shahrooz Elect Res Co, R&D Managment	3	0	0
Glicksman, R.L.	2	USA	George Washington Univ	24	319	10

TABLE 11.2 Main authors in wastewater taxes research (*cont.*)

Authors	Articles	Country	Affiliation	Total publications	Total citations	h-Index
Hong, J.L.	2	China	Shandong Univ	92	1,807	24
Hong, J.M.	2	China	Liaoning Shihua Univ	14	357	7
Marufuzzaman, M.	2	USA	Mississippi State Univ	45	667	15
Mueller, J.F.	2	Australia	Univ Queensland	490	12,815	60

SOURCE: OWN ELABORATION BASED ON WOS (2020)

are five scientists with more than 100 publications (their total research) and an h-Index over 20; namely, Sánchez-Pérez J. A. (Spain), Molinos-Senante, M. (Chile), El-Halwagi, M. M. (USA), Hong, J. L. (China) and Mueller J. F. (Australia).

Figure 11.6 presents the main institutions focusing on wastewater research. These 18 institutions account for 47.92% of the total articles, which indicates a high concentration index at the institutional level. The leading institution in number of related articles is the University of Almeria (Spain), with 7 articles

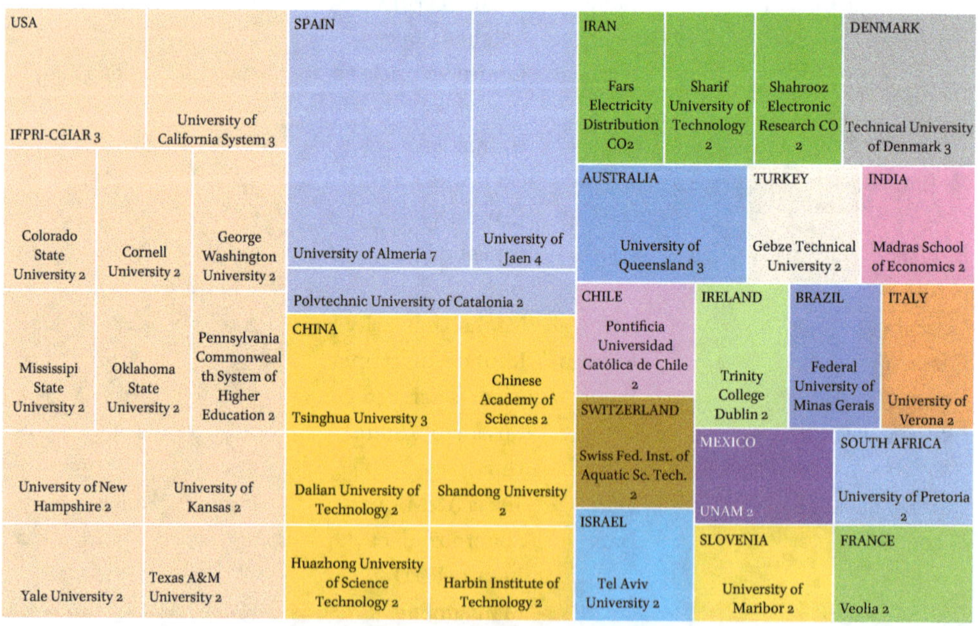

FIGURE 11.6 Main institutions in wastewater taxes research
SOURCE: OWN ELABORATION BASED ON WOS (2020)

(4.86% of the total), followed by the University of Jaen (Spain), with 4 articles (2.78% of the total). There are then quite a few institutions with 3 results each (2.08%): IFPRI-CGIAR (USA), University of California System (USA), Technical University of Denmark (Denmark), Tsinghua University (China) and University of Queensland (Australia). The main institutions showed in Figure 11.6 together account for 61% of the total results, illustrating the high dispersion of articles among different bodies.

4 Discussion and Conclusions

At a global level, research on wastewater taxes has only yielded 144 publications. This subject remains markedly understudied compared to other environmental taxes, accounting for a very small share of environmental taxes research in the period 2000–2020 (1.39%). In comparison, electricity taxes, gas emission taxes and waste taxes have attracted far more attention, with shares of 20.71%, 20.56% and 15.41%, respectively. The bibliometric analysis carried out reveals a subject that calls for further investigation, given the scarcity of scientific publications in this area. It is worth noting, however, the relevance of the Spanish contribution to this small body of research, as it represents 11.11% of the total. Spain is among the countries that has produced the most research on this subject during the period 2000–2020, along with the USA (25%) and China (14.58%). These leading countries are followed by Italy (5.56%), Australia, Brazil, Denmark, Germany, India, and Iran (2.78% each) and Ireland, the Netherlands, Norway, Switzerland, and Taiwan (2.08% each).

The scientific output on this subject has gradually increased during the period under study, although with some fluctuations, registering a marked rise from 2010 on. This growth corresponds to greater concern about ensuring environmental sustainability, as reflected in the United Nations Millennium Development Goals adopted in September 2000 (UNDP 2020) and the implementation of certain regulations; for example, in the European Union, the Water Framework Directive 2000/60/EC and Directive 91/271/EEC have helped drive European policies guaranteeing the quality of water resources. Taxes as a tool to tackle pollution are justified by the "polluter pays" principle (Pigou 1920), which seeks to correct the negative externality caused by pollution. The idea is that every market agent should pay for the pollution they generate, encouraging the search for less polluting profitable activities as well as investment in cleaner technologies. In terms of scientific production at institutional level, the leading Spanish institution in wastewater taxes research is the University of Almeria, which has published articles in several notable high impact journals, such as Water Science and Technology, Desalination, Science of the Total Environment and Desalination, with a JIF of 1.638, 6.603, 6.551 and 7.098, respectively. In addition to these journals, there are many more with high index scores that have published papers on this subject, including journals in the top two quartiles of the SJR such as Waste Management, Journal of Cleaner Production, Resources Conservation and Recycling, Water, Agricultural Water Management, Applied Energy, Bioresource Technology, Desalination and Water Treatment, Energies, Environment International, Industrial & Engineering Chemistry Research, International Journal of Life Cycle Assessment, Journal of Environmental Management and Sustainability.

Fiscal instruments are necessary for the proper management of resources; in this case, they are particularly important since access to water is essential for life. The implementation of wastewater taxes fulfils three basic functions according to Riera (2005): (i) it frees up public resources that can be directed towards other purposes; (ii) it promotes greater business efficiency by linking the revenue collected to the service provided and (iii) it creates a clear signal for users about the real cost of the service, encouraging rational consumption. In addition, the revenue generated by taxes can help finance reclamation systems, which increase the supply of water available, especially for irrigation. Such taxes can also positively contribute to reducing the discharge of untreated effluents, preventing harmful effects on the environment. This study serves as a guide for researchers, pointing out where new trends may emerge, and showing that there is a segment that is still to be explored. The improvement in the management of environmental taxes necessarily implies a growing interest in the subject by all sectors involved, and in particular by the scientific community and institutions. National governments need support to explore the different options available to them for funding wastewater management, such as charging water/wastewater taxes. In addition, institutions should make data and information publicly available in sufficient quantity and quality to be used in research, as this sector is characterised by a transparency gap, complicating the task of accessing information useful for research and consequently making it more difficult to improve management in this area. As well as enabling research, said information can be used to evaluate the actions carried out and carry out dissemination actions to raise public awareness about protecting the natural environment and the need for wastewater taxes to improve management.

References

Archambault, É. D., Y. Campbell, and V. Gingras. 2009. "Comparing bibliometric statistics obtained from the Web of Science and Scopus." *Journal of the Association for Information Science and Technology* 60 (7): 1320–6. https://doi.org/10.1002/asi.21062.

Aznar-Sánchez, J. A., J. J. García-Gómez, J. F. Velasco-Muñoz, and A. Carretero-Gómez. 2018. "Mining Waste and Its Sustainable Management: Advances in Worldwide Research." *Minerals* 8 (284). https://doi.org/10.3390/min8070284.

Bluffstone, R. A. 2003. "Environmental taxes in developing and transition economies." *Public Finance and Management* 1. https://doi.org/10.2139/ssrn.461539.

Bryman, A. and E. Bell. 2015. *Business Research Methods.* USA: Oxford University Press.

Burak, S., and J. Margat. 2016. "Water Management in the Mediterranean Region: Concepts and Policies." *Water Resources Management* 30 (15): 5779–97. https://doi.org/10.1007/s11269-016-1389-4.

Connor, R., A. Renata, C. Ortigara, E. Koncagül, S. Uhlenbrook, B. M. Lamizana-Diallo, and D. Brdjanovic. 2017. *The United Nations world water development report 2017. Wastewater: the untapped resource.* UN-Water: UNESCO.

Durieux, V., and P. Gevenois. 2010. "Bibliometric indicators: Quality measurements of scientific publication." *Radiology* 255: 342–51. https://doi.org/10.1148/radiol.09090626.

Falagas, M. E., V. D. Kouranos, R. Arencibia-Jorge, and D. E. Karageorgopoulos. 2008. "Comparison of SCImago journal rank indicator with journal impact factor." *The FASEB Journal* 22 (8): 2623–8. https://doi.org/10.1096/fj.08-107938.

Gallego-Valero, L., E. Moral-Pajares, and I. M. Román-Sánchez. 2018. "The tax burden on wastewater and the protection of water ecosystems in EU countries." *Sustainability* 10 (1), 212. https://doi.org/10.3390/su10010212.

Huang, L. Y., Y. Zhang, Y. Guo, D. Zhu, and A. L. Porter. 2014. "Four dimensional Science and Technology planning: A new approach based on bibliometrics and technology roadmapping." *Technological Forecasting and Social Change* 81: 39–48. https://doi.org/10.1016/j.techfore.2012.09.010.

López-Serrano, M. J., J. F. Velasco-Muñoz, J. A. Aznar-Sánchez, and I. M. Román-Sánchez. 2020. "Sustainable Use of Wastewater in Agriculture: A Bibliometric Analysis of Worldwide Research." *Sustainability* 12 (21), 8948. https://doi.org/10.3390/su12218948.

Montoya, F. G., R. Baños, J. E. Meroño, and F. Manzano-Agugliaro. 2016. "The research of water use in Spain." *Journal of Cleaner Production* 112: 4719–32. https://doi.org/10.1016/j.jclepro.2015.06.042.

Pérez, A. T. E., M. Camargo, P. C. N. Rincón, and M. A. Marchant. 2017. "Key challenges and requirements for sustainable and industrialised biorefinery supply chain design and management: A bibliographic analysis." *Reviews* 69: 350–9. https://doi.org/10.1016/j.rser.2016.11.084.

Pigou A. 1920. *The Economics of Welfare.* Accessed 5 December 2020. http://www.econlib.org/library/NPDBooks/Pigou/pgEW0.html.

Prigle, J. 2008 "Trends in the use of ISI citation databases for evaluation." *Learned Publishing* 21 (2): 85–91. https://doi.org/10.1087/095315108X288901.

Riera P. 2005. *Manual de Economía Ambiental y de los Recursos Naturales.* Spain: Paraninfo.

Romo-Fernández, L. M., V. P. Guerrero-Bote, and F. Moya-Anegón. 2013. "Analysis of the Spanish scientific production in Renewable Energy, Sustainability and the Environment (Scopus, 2003–2009) in the global context." *Investigación Bibliotecológica* 27 (60): 125–151. https://doi.org/10.1016/S0187-358X(13)72546-2.

Trinh, L. T., C. C. Duong, P. Van Der Steen, and P. N. Lens. 2013. "Exploring the potential for wastewater reuse in agriculture as a climate change adaptation measure for Can Tho City, Vietnam." *Agricultural Water Management* 128: 43–54. https://doi.org/10.1016/j.agwat.2013.06.003.

Uche, J., A. Martínez-Gracia, F. Círez, and U. Carmona. 2015. "Environmental impact of water supply and water use in a Mediterranean water stressed region." *Journal of Cleaner Production* 88: 196–204. https://doi.org/10.1016/j.jclepro.2014.04.076.

UNDP. 2020. *Millenium Development Goals.* Accessed 5 December 2020. https://www.undp.org/content/undp/en/home/sdgoverview/mdg_goals.html.

UN-Water. 2020. *United Nations World Water Development Report 2020: Water and Climate Change.* UN-Water: UNESCO.

Vázquez, V. L., G. Rodríguez, T. Daddi, M. R. De Giacomo, C. Polders, and E. Dils. 2015. "Policy challenges in transferring the integrated pollution prevention and control approach to Southern Mediterranean countries: a case study." *Journal of Cleaner Production* 107: 486–97. https://doi.org/10.1016/j.jclepro.2014.06.090.

Vörösmarty, C. J., P. Green, J. Salisbury, and R. B. Lammers. 2000. "Global water resources: vulnerability from climate change and population growth." *Science* 2000 289 (5477): 284–8. https://doi.org/10.1126/science.289.5477.284.

Weerasekara, P. 2017 "The United Nations World Water Development Report 2017 Wastewater." *Future of Food: Journal on Food, Agriculture and Society* 5 (2): 80–81.

Zárate Marco, A., J. Vallés Giménez, and C. Trueba Cortés. 2007. "Descentralización fiscal y tributación ambiental: el caso del agua en España." *Papeles de trabajo del Instituto de Estudios Fiscales. Serie economía* 24: 1–55.

CHAPTER 12

The Symbolic Configuration of Andalusian Rivers: Iconography and Artistic Values

Antonio Rafael Fernández-Paradas and Rubén Sánchez Guzmán

Abstract

The objective of this research is to identify, analyse and contextualise the phenomenon of the identification of Andalusian rivers with allegories. It is a national and international phenomenon, which developed a series of iconographic particularities in Andalusia. Likewise, it is an issue that extended until the twentieth century.

Keywords

allegory – iconography – river – symbolism – water

1 Introduction

Quite a lot of societies became aware of the importance of their rivers as a life-sustaining element of their civilisation throughout History. The most paradigmatic case may have happened with the Nile River (Gómez 2014–2015 and 2018), whose banks gave life to one of the most important civilisations, the Antique Egypt. During millennia and with extraordinary precision, people had to measure and control the Nile's floods as they wished and needed. Hence, it does not come as a surprise that a society (Egyptian or otherwise) begins to deify these water flows, since they base a large part of their economy and subsistence on the rise and fall of the waters and what could be done with them. The next step would be the material embodiment in painting, sculpture, ceramics, stone, etc., of those rivers that were to be worshipped (Hesíodo 1978).

The deification of the river Nile (Jentel 1987; Swetnam-Burland 2009) was matched by a multitude of national rivers that marked the character of their peoples, such as the Eurotas river (Brewster 1997), the Tiber, or the Ebro. This phenomenon must be understood as a process of settlement of the Greco-Latin culture (Torres 1990), where the different territories followed

 | DOI:10.1163/9789004679634_013

the example of what was happening in other parts of the empire, especially in Rome. The Andalusian context was no stranger to this phenomenon either: its two great rivers, the Guadiana, and the Guadalquivir, received the greatest attention from artists. Most of the examples that we have been able to verify from the classical period (Costa, Palahí, and Vivo 2011) are related to sculptural and numismatic representations of the Guadiana River. It has not been possibly to find any representations of Andalusian rivers during the medieval period. During the modern period, until the twentieth century, has been confirmed the supremacy of the representation of the allegory of the Guadalquivir River. This one is the Andalusian river most frequently depicted in engravings, sculptures and paintings. It should be noted that this is a river that runs through a large part of Andalusian territory, passing through important capitals such as Cordoba and Seville, and a large part of its course was navigable during the modern period.

The representations found of other Andalusian rivers are practically non-existent, being reduced to the river Guadalete and the Darro in Granada. In 1929 there is a representation of the Guadalimar as a tributary of the Guadalquivir. The twenty-first century saw representations of the Genil and the Tinto and others of the Guadalquivir and the Guadiana.[1] It is striking, due to their non-existence or lack of knowledge, that there is no representation of paradigmatic rivers such as the Guadalhorce and Guadalmedina in Malaga, the source of so much wealth, but also of so much misfortune. This may be because of the interest that the city has always had in the sea, and not in its rivers (Montero 2012).

Gómez Mayordomo has defined the personification of the rivers in a representative way, mentioning that "in general, the river-gods were represented with their torso naked and dressed only in a cloak that covered their legs and hips. They also wore headdresses with lake leaves and carried a reed branch, an oar, and/or a cornucopia, symbol of their fertility. Finally, they were sometimes shown resting their arm on a pitcher from which water flowed" (Gómez 2014–2015, 88. This is a definition that fits, with minor nuances, perfectly with the presentation of Andalusian rivers from classical antiquity to the twenty-first century).

Finally, it is necessary to mention that the "canonical" representation (Ostrowski 1999) of the iconographic model means that some preserved examples can be identified as the specific representation of a river, being "abstract" allegories of rivers (Gais 1978; San Nicolás 1997 and 2004–2005).

1 The rivers mentioned here can never be understood as definitive. It is a catalogue of works in the process of revision and growth.

2 The First Representations of the Iconographic Model and Its Consolidation in the Classical Period

The importance of rivers gave rise in Greece to the mythological creation of the three thousand Oceanides who governed the river currents of the known world in Antiquity. According to Hesiod, they were sons of two primordial deities (the Titans Oceanus and Thetis) and siblings of three thousand Oceanides, or nymphs (Hesiod 1978, 85–6). Some of them were worshipped despite being minor deities, given their life-giving and fertilising character, associated with water. Classical texts spoke of the personification of these rivers, who also took part in the narrative of the myths. However, the iconographic representation was not stable at first, and was gradually modulated until it was finally codified. The first plastic personifications of rivers date back to the seventh century BC. The myth of the river god Achelous (who fought with Herakles for Deyanira) set the tone; in the beginning, his image was a hybrid between serpent and human (iconography borrowed from the god Oceanus); his representation was quickly replaced by the figure of a bull (an image representing fertility), or by its anthropomorphic variant, a hybrid between bull and human; later we can find the iconography of the androsopos bull: a bull with the face of a man (Ciuccarelli 2006). Although this personification was exclusive to the Aqueloo river, it ended up contaminating other iconographies. In the second half of the fifth century the iconography of rivers began to shift towards increasingly humanised forms, first retaining certain minor animalistic residues (such as horns and ears), and then moving on to fully anthropomorphic forms. The origin of the humanised image of the river is unclear: while some authors believe that it can be found in the pediments of the Parthenon and the temple of Zeus at Olympia, i.e., at the height of the classical period (Ostrowski 1991, 21), others date it to the Hellenistic period (Gais 1978, 361). However, the iconography at that time was not uniform either: for in addition to the reclining god, and its seated or leaning variants, there were others (the rising river, widely represented on coinage and which seems to derive from the sculpture of the Antiochian Tychic or Fortuna with the river Orontes at her feet, carved by the sculptor of the Alexandrian focus Eutychides around 300 BC, known from Roman copies). However, it would be the aforementioned personification of the river as a reclining man the one that would become popular in the Hellenistic period and enjoy greater importance later on, practically up to the present day. A key piece for its later development was the Alexandrian figure of the Nile, well known from the copies of the Roman period, such as the piece in the Vatican Museums. In essence, these images show mature men in a reclining position, resting their arm on a kind of urn from which water flows as a reflection of its flow, showing their naked torso, covering their hips and legs with a cloak. They are generally bearded (large rivers of the empire), but other smaller rivers may appear bearded. They are adorned with aquatic plants, and as attributes they usually carry cornucopias, as a symbol of fertility, or oars, due to their navigable condition. However, the symbolic values given by the Romans to these deities would transcend the merely religious Greek ones, to give them other values such as propagandistic ones on coins, columns, and triumphal arches, or allusive to the local geography, associating these images with the personification of the cities themselves.

Roman Hispania was no stranger to this trend, although there are few reliable examples of representations of rivers in the Iberian Peninsula and more specifically in Andalusia, and they can sometimes create confusion with other iconographic representations. The presence of rivers is frequent in mosaics, but they are foreign rivers (Asopo, Aqueloo, Euphrates, Nile, Orontes, Pyramus) in most of the cases identified, associated with a narrative scene of a mythological nature or in isolation, facilitating their identification with a "tituli" or inscription nearby, or with some identifying attribute (San Nicolas 2004–2005, 301–33). In Baetica we only find perfectly identified representations of the rivers Aqueloo, Nile, Asopo and Tiber. More doubtful is the central image of the mosaic in the house of the planetarium of Italica, which could be the Betis, or the Aquello (Granados 2019, 53), who has such a generic iconography (an old man, reclining, and carrying a body of plenty), that it is impossible to identify him with certainty. The same iconographic insecurities are found in the sculptures of certain Betic river deities. There are sculptures that may have been associated with the decoration of fountains (Loza 1993b, 99) in the Archaeological Museum of Cordoba,[2] and the Archaeological Museum of Seville. Both show the same formal characteristics (a male figure reclining over a fountain, covering his lower limbs with a cloak). However, as they do not bear any kind of inscription, and lack any specific attribute, their identification with a specific river is highly risky. In the case of the piece in the Cordoba Museum, its beardless features would prevent it from being identified as a great river of the magnitude of the Betis, so it may be one of its tributaries (Loza 1993a, 146). Meanwhile, the Sevillian figure, from a patrician dwelling in Italica, has a similar iconography, as he is shown reclining on a cantharus that serves as a water channel, the only identifying element, as he is still headless and without a right arm (León 1995, 54).

2 Inventory no.: CE028883.S

FIGURE 12.1 Anonymous. Lintel of the rivers. Ana and Albarregas rivers
SOURCE: MUSEUM OF ROMAN ART OF MERIDA, MERIDA

In nearby Lusitania, watered by the *Flumen Ana* (now called Guadiana), there is a similar phenomenology around its capital Augusta Emerita, where everything seems to indicate that the waters of the river were worshipped according to the pieces preserved. An exceptional work of capital importance, is the so-called Lintel of the Rivers from the Museum of Roman Art in Mérida[3] (Figure 12.1), which has wo fluvial deities identified with inscriptions and comes from the funerary monument of Caius Julius Successinus dated to the third century (Canto, Bjearano, and Palma 1997, 247–94). In it, two rivers are clearly identified with silk inscriptions. The *Flumen Ana*, is placed on the left, together with the inscription *B. B.* interpreted as Bonis Bene, in reference to the prosperity and fertility brought by the river, or by Successinus himself, and its tributary at Barraeca, today the Albarregas, on the right. Both have similar formal characteristics as they are represented leaning on an amphora from which the water flows, and carrying reed sticks, with the only difference being the Guadiana, which is bearded, assuming the iconography of other great rivers, with the Albarregas depicted without a beard, as it is considered a tributary or minor river.

Related to this fluvial cult is the sculpture discovered in 1902 in excavations on the hill of San Albín, from the Mithreo of Mérida, now in the National Museum of Roman Art.[4] Similar in model to the fluvial deities present in the cosmological mosaic from Emerita, the headless male figure lies on a watery bed with incised waves, sitting on a dolphin, while with the missing right arm he carries a cornucopia. Given the generality of these attributes, different interpretations were put forward from the outset, ranging from identification with the Titan Oceanus, or with the Flumen Ana, a much more plausible and generally accepted theory (Nogales 2002, 99–101).

In view of the above examples, it is not surprising that Hispanic Roman numismatics, so sparing in fluvial personifications as opposed to what

3 Inventory no.: DO2012/2/3.
4 Inventory no.: CE00085.

happened in other imperial mints, has an exception in Augusta Emerita. There are a series of coins, minted in the city in the time of Augustus, with the image of a bearded old man with an amphora from which water flows before his mouth. This was traditionally considered as an image of Silenus, but taking into account the pieces described above, the theory that it was a personification of the Guadiana has gained strength (Álvarez 2020, 18–20).

3 The Dissemination of the Model through Engraved Sources: Renaissance Contributions

Although the Middle Ages did not completely obliterate personifications, and the river's iconography remained latent, the rivers represented were adapted to particularities of the Christian faith, such as the river Jordan (Rodríguez 2002, 289) or the rivers of Paradise, in all kinds of techniques and supports (Gómez 2019, 55–86). However, with the arrival of the Renaissance and the return to Roman models, the classical models and their use resurfaced. The first factor to be taken into account at this time was the discovery and study of classical sculptures. From the early sixteenth century onwards, excavations began to uncover new sculptures which, together with those already known, formed a powerful group of homogeneous iconographies. In 1512 the monumental sculpture of the river Tiber from the Iseo Campenseo (a sanctuary dedicated to the deities Isis and Serapide), was discovered in the Marzio field (now it is in the Louvre Museum, after belonging to the Vatican collections until 1804, when it was seized by Napoleon). In 1513, and close to the previous discovery, the image of the Nile in the Vatican Museums was unearthed, a Roman copy of an Alexandrian original described by Pliny the Elder in the interior of the Forum of Peace. Perfectly individualised, the Tiber is accompanied by Romulus and Remus, and the Nile does so surrounded by pygmies and leaning on a sphinx.

Two other Roman pieces were placed in the sixteenth century in the populous Capitoline square in Rome, presiding over the fountain on the main façade of the senatorial palace. Although today they represent the rivers Tiber and Nile, the first one (at least), has undergone a re-education: the tiger's head on its right arm has been replaced by a small sculptural group of twins next to the head of the Capitoline she-wolf, adjusting it to the geographical reality of the city of Rome. It goes without saying that this was not an isolated phenomenon and the appearance of these specimens aroused enormous interest, becoming true iconographic types. As a result of these findings, some statues that had been preserved in fragments were reintegrated and served as models (for example, the fluvial divinity identified as the river Arno in the Vatican).

From the time they were discovered, these sculptures enjoyed enormous popularity as they were engraved, which ensured that they were widely distributed beyond the frontiers of Rome. The engraver Nicolas Beatrizet, French by birth but settled in Rome, where he worked with Michelangelo, made engravings of the pair of rivers of the Capitoline Hill raised on their pedestals, as well as the Tiber in the Louvre and the Nile in the Vatican, These were compiled, along with many others, in the *Speculum Romanae Magnificentiae*, published by Antonio Lafreri, which underwent successive copies and variants, reflecting the enormous popularity and diffusion of the same, from its first editions in the decade of the seventies of the sixteenth century. In addition to these examples, which are direct interpretations of the aforementioned sculptural models, other individual prints were added that included the classical iconography, not only in mythological scenes, but also in family exaltations (such as the one produced in 1589 by Cherubino Alberti of the Medici Family), or alone, identifying themselves with the different cities in where they bathe. Throughout the sixteenth century there was a recoding of the specific attributes of these river deities, maintaining those created in classical antiquity and incorporating new ones. A good example of this was the publication in Rome in 1593 of Cesare Ripa's Iconology. This compendium of emblems and allegories was no stranger to the personifications of rivers, although it is the version of antiquity through the lenses of the Renaissance. There, Ripa personalises the iconographies of the great rivers of the Roman Empire, to which new ones are added: from Asia and Africa, the main rivers of the Italian peninsula, the rivers alluding to classical mythology (just two taken from Ovid's Metamorphoses), and the rivers of the underworld taken from the works of both Virgil and Dante. The omission of other European rivers outside the Italian peninsula (except for the Danube) is total. All this conditioned the arrival of the iconography of the rivers to other parts of Europe: in Spain and Andalusia they did so with the absence of local iconographies, so it was necessary to borrow the iconographic models created since the sixteenth century in Italy, which were spread throughout Europe, and which were also a continuation of the classical types (Fernández-Paradas and Sánchez-Guzmán, 2021).

4 Works and Iconographies from the Sixteenth to the Twenty-First Century

There are two engravings of the river Guadalquivir dated on the sixteenth century, and another two from the second half of that century (one from 1570 and the other from 1588). Sculptures or paintings have not been found.

Although these are only two examples, they are two pieces that provide some of the few novelties that we can see in the evolution of river's iconography in Andalusia. The first of these is an allegory of the river Betis which illustrates the text by Juan de Mal Lara, about the reception given by the city of Seville to Philip II in 1570 (De Mal [1570] 1992, 99–101). The second is an engraved map of the kingdom of Jaen, published in 1588.[5]

With regard to the representation for the entry of Philip II in 1570, we should point out that this is the only representation of an Andalusian river until the twentieth century (or almost) that we have been able to find. In it, the figure appears standing, this pose being explained by Betis coming "back, with the crescent, as ambassador to His Majesty, from the Ocean, father of all waters". He is a person of mature age, partially covered with a cloak that covers the genital area in particular. He has a crown of reeds, olive trees and cattails, and he holds a pitcher from which water is pouring. In the lower part, an urn spouting water, supports the left foot, and next to it we can see a dolphin and two swans. We can also read verses alluding to the reverence of the river and the ocean itself to Philip II. The format was probably intended to be more functional than its transfer to ephemeral architecture. Except for the verticality of the composition, it is a representation that conforms to traditional modes (Morillas 1991).

The second verified work from the sixteenth century is a map of the kingdom of Jaen (Figure 12.2). It depicts the different towns of the kingdom spread throughout the territory like a tangle whose axes are perfectly coded by the omnipresence of the Guadalquivir River and all its tributaries. The Guadalquivir River has its source in a double branch that extends to the lower right margin of the map, where we can see a portentous figure represented as a naked old man, without a crown, and with a pitcher letting water out, and two important inscriptions, "Betis" and "Tartesio" (referring to the Latin name of the river, and also to the Roman province and the Tartessian civilisation, probably placed on the map to emphasise the antiquity of the kingdom of Jaen). The representation leaves no room for doubt: life in the kingdom of Jaen is developed thanks to the river-god Guadalquivir. The double source of the river is located in Cazorla and Quesada. The figure of the god is ornamented with two other attributes: a plant rod and a horn of plenty from which fruits and plant stems emerge. The main tributary of the River Guadalquivir, the River Guadalimar, is born from the horn of plenty, once again emphasising the importance of the Guadalquivir and the river-god as a source of life for the kingdom.

5 National Library (Madrid) – M. 8v. Document number: IECA1988015105. CDU number: (468.14).

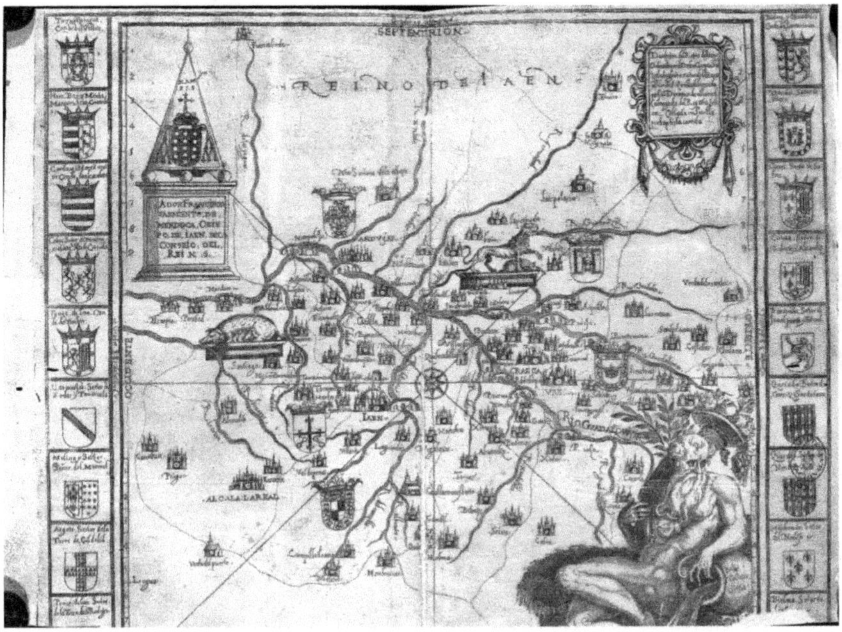

FIGURE 12.2 Gaspar Salcedo de Aguirre. Map of the kingdom of Jaen
SOURCE: JUNTA DE ANDALUCÍA, INSTITUTE OF STATISTICS AND CARTOGRAPHY

The seventeenth century was a poor century (or one with few surviving testimonies) in terms of the personification of Andalusian rivers. Note one work, an important example due to the way in which the iconography is resolved. It is a canvas by the Granada painter Pedro Anastasio Bocanegra (Figure 12.3) and is kept in the Museum of Fine Arts in Granada[6] . The work presents two important novelties: firstly, there is a young blond, beardless man (not a bearded old one), and he carries as an attribute a sceptre ended in a granada fruit. The young man is crowned by two angels. This is an unusual work in the painter's career. An engraving by León Davent depicting Diana the Huntress at rest has been cited as the source of the painting. It is considered that, since it is a tributary of the Genil, the iconography follows the classical canon in depicting tributaries much younger than the main river. Two allegories are attributed to Bocanegra, one of Justice in the Royal Academy of Fine Arts of San Fernando and other one of the Plague in the Goya Museum in Castres. One might

6 Inventory no.: DE119.

FIGURE 12.3 Pedro Anastasio Bocanegra. Allegory of the River Darro. Circa 1676–1680
SOURCE: MUSEUM OF FINE ARTS, GRANADA

wonder whether this representation of the Darro could have been originally paired with a work depicting the Genil.

The scarcity of representations of rivers in the seventeenth century extends to the middle of the eighteenth century. From the second half of the eighteenth century, there are several representations of allegories of the river Guadalquivir (five) and an emblem of the river Guadalete. The La Granja fountain (Herrero 1995) is also part of this context, with the representation of several rivers, presumably including several Andalusian ones, but which are difficult to identify.

The six analysed representations that depict the Guadalquivir River correspond to an engraving made in 1747 as the title page of Lorenzo de Zúñiga's *Anales Religiosos y Seculares* (Morillas 1991, 305–322; Zúñiga 1671): first, a mid-eighteenth century canvas by Juan de Espinal in the Real Academia de Bellas Artes de San Fernando[7] entitled "Llegada de la pintura a Sevilla/, Alegoría de la pintura sevillana" ("The Arrival of Painting in Seville/, Allegory of Sevillian Painting") (Pérez 2012; Quiles 2007, 104–5, 114, 107–132; Santiago and Cera 2016; Portús 2016, 224); second, another anonymous one in the Museum of Fine Arts of Cordoba, of an allegory of the city through its numerous illustrious sons (in this one the personification of the river appears, together with another figure, bearing the coat of arms of the city[8]), third and fourth, two descriptions of the river Guadalquivir with an allegorical representation each, one preserved in

7 Inventory no.: 0131.
8 Inventory no.: CE2314P.

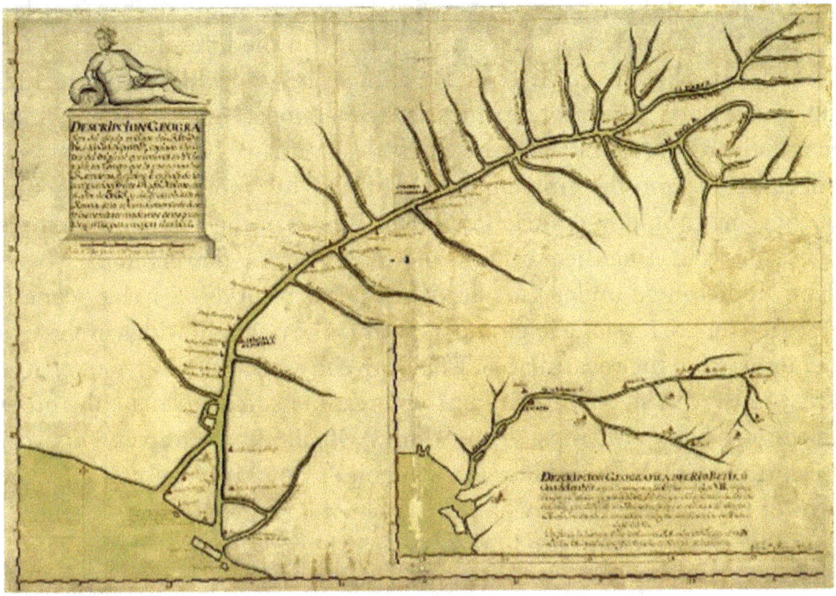

FIGURE 12.4 Anonymous. Allegory of the River Guadalquivir. Circa 1770
SOURCE: MADRID. ARMY GEOGRAPHICAL CENTRE

the Centro Geográfico del Ejército (Madrid) (Figure 12.4),[9] fifth, other in the Museo Marítimo Torre del Oro in Seville,[10] whose figures seem to evoke the aforementioned engravings by Beatrizet, of the rivers of the Roman square of the Capitol; and finally, a medal from 1778 with a representation of the river is kept in the Museo Cerralbo in Madrid.[11]

In all cases, these representations conform to the traditional canonical model (i.e., a reclining, naked, elderly, bearded man, with or without a crown). Among the examples mentioned, the one with the greatest iconographic richness is the Cerralbo Museum medal made in 1778. It shows the river Guadalquivir lying down, looking to the right, leaning on a pitcher from which the river rises, while with his left hand he holds a rudder (it is a navigable river). In the background there are two sailing boats. On the reverse we can read the inscription DIES ET INGENIUM. Together with two spinners, a carder, and a weaver with tools for carding, spinning, and weaving.

Within this group of eighteenth-century works, the emblem of Jerez de la Frontera and its idealisation of the river Guadalete is a separate case. It exalts

9 Army Geographical Centre (Madrid). Arm. G. TBLA. 5th Carpt. Iª no. 19. ICA1988000190.
10 Torre del Oro Maritime Museum (Seville). 170.ICA1988000861.
11 Inventory no.: 02884.

the city of Jerez and its river as a paradise. Moreno Arana describes it as follows: "... in the third part he (the artist) deals with the interpretation of the emblem or hieroglyphic that presided over the class, and which gives the title to the whole work. Thus, as one entered the classroom, one could see on its main wall the royal coat of arms on which stood Mercury's caduceus as a symbol, deciphered from the emblem CXVIII of Alciatus' repertory, of "wisdom, eloquence and promptness in action". At both sides of the coat of arms there were two flasks of Amalthea or horns of plenty pouring flowers to the young students, represented under the allegory of flocks of birds (...). This scene is represented on the Elysian Fields and its river Lethe (...) from which emerges a border embracing the royal coat of arms with the epigraph "STAT PROVIDUS, OBVIUS, PERVIUS", signifying firmness and readiness to act (STAT), the other three epithets being "the most visible things, which rather intervene in every praiseworthy work". The emblem (...) symbolises how the wise and effective decrees of the government of Charles III, which have given rise to "the most beautiful flowers of eloquence in the Masters", are poured over the Jerez elysium so that the living souls or "manes" that dwell in them, who are none other than their students, find "the intellectual nourishment of the soul, the beautiful letters" (Moreno 2014, 162–3).

4.1 *The Issue of the Representations of the Andalusian Rivers at La Granja*

The iconographic programme of the gardens and fountains of La Granja de San Ildefonso would be the first iconographic programme in which the main rivers of Spain appear (since in the previous artworks the rivers are of a local nature). There are all kinds of myths of antiquity and allegories (designed after the Versailles model) in the decorative orchard of the royal gardens, and with those myths and allegories we can see both the geographical dimension of the kingdoms and the power of the monarchs who financed these works reflected through allegories of the different rivers that irrigated their vast territories (these rivers being a source of wealth and abundance in their respective kingdoms). In Louis XIV's Versailles, the Loire, the Rhone, the Seine, and the Garonne are represented as venerable old men; their respective tributaries, the Loiret, the Saône, the Marne and the Dordogne, are transformed into graceful maidens in the ponds of the Water Parterre. In Philip V's Granja, the rivers of the Iberian Peninsula appear in a similar dual iconography, masculine and feminine. However, unlike the French case, and due to the laconic nature of the written sources, there is no unanimity in their identification, as there is no specific iconographic model. The historiography is frankly confusing when talking about the Guadalquivir: for some experts, the Guadalquivir is the venerable old man who pours his waters into the pond of the new waterfall (Martín 1825, 25), for others it is one of the two maidens or nymphs who pour their waters into the pond from the base of the central rock of the Fuente de la Fama fountain. In addition to the disparate criteria of the written sources, they lack any identifying icnographic element that helps identify them. The aforementioned female images (if we follow the Versaillesque model, they would be female allegories of rivers of lower rank), have been seen by some writers as allegories of the Guadalquivir and the Guadiana, but without individualising the identity of each; to some, the Pisuerga replaces the Guadalquivir (Martín 1825, 59). However, for Ponz, the four main rivers of Spain would be the Tagus, Duero, Guadalquivir, and Ebro (Ponz 1787, 172), leaving out the Guadiana. In the case of the maiden flanking the Amphitrite Pond of the new cascade on the right, paired with another river effigied as a venerable old man, the sources agree that it would be the personification of the Guadiana River (Ponz 1787, 163; Martín 1825, 25 and 163), as a nymph reclining on a large jar like a fountain, accompanied by a group of children playing with a fish and a tortoise. Its female representation would be explained by the fact that it is an important but minor river, and because the traditional iconography of a venerable old man is usually reserved for other rivers in the Iberian Peninsula, such as the Ebro, Tagus, Duero, etc.

4.2 *Revisions of Iconography in the Nineteenth Century*

Two other works from the nineteenth century are worthy of being talked about: a painting and a drawing, featuring the Andalusian river. The first is a painting by Francisco Aznar García for the Conference Hall of the Congress of Deputies in 1861, a space decorated with Neoplateresque stucco paintings and mouldings. Allegories of the continents were depicted on the glazed vault, while on the walls are twelve paintings with themes relating to the ancient kingdoms of Spain, the provinces, and the cities. The four main rivers of the kingdom, the Ebro, the Tagus, the Duero, and the Guadalquivir, are also depicted (Salvá 1997).

The representation of the river presents some novelties when compared to the traditional configuration of the image. The pitcher from which water gushes and the oar, found for example in the medal in the Museo Cerralbo, are retained, but now we see an adult man, not an old one, not bearded, crowned and almost completely naked, except for a small cloth covering the genital area. In his hand he holds a bouquet of ears of corn, fruit, and leaves.

The second work to be talked about is a drawing from 1892 by Ángel María Barcía Pavón in the Museo de Bellas Artes in Cordoba (Figure 12.5).[12] According

12 Inventory: CE1883D.

FIGURE 12.5 Ángel María Barcía Pavón. Allegory of Cordoba and the Guadalquivir. 1892
SOURCE: CORDOBA MUSEUM OF FINE ARTS, CORDOBA

to the Cordoba Museum, the work was part of the Collection that Ángel Avilés donated to this museum on his death. The work was executed in 1892 with a view to being reproduced in a magazine whose subject would allude to the floods caused by the Guadalquivir as it passed through the city that year, as can be deduced from various literary testimonies and from the drawing on the same subject which is also kept in this Museum from the Avilés Collection, by Rafael Hidalgo de Caviedes. The composition depicts an angel, most probably

Saint Raphael, resting on the figure of a bearded old man pouring a pitcher of water, symbolising the river Guadalquivir. In the upper area and in different planes, three representative architectures of monumental Cordoba can be seen: the Gate of the Bridge, the tower of the Cathedral and the Triumph of Saint Raphael. At the bottom, a general view of Cordoba from the Campo de la Verdad.

Regardless of the identification of the angel with the archangel Saint Raphael, patron, and protector of the city, it is interesting to note the moralisation of classical iconography under the protection of Christianity. The angel's intercession is a warning to the river-god for him to stop flooding the city. The angel firmly grasps the god's arm, while the god looks at him sternly. As for the iconographic representation of the river, it is depicted according to the norm.

4.3 *Perpetuating an Iconographic Model*

After the disaster of' 98, Spain entered the twentieth century with an identity revisionism that deeply marked Spanish society at the time. The country saw the need to redefine itself and find itself again within the international concert. This process culminating, quite a few years later, with the holding in 1929 of two important events on an international scale: the International Exposition in Barcelona and the Ibero-American Exposition in Seville.

The Barcelona International Exhibition was the ideal occasion to reflect, through a monumental fountain placed in the centre of the square of the same name (the heart of the Montjuic exhibition site), the image of the Spain that was to be transmitted. This was done with a great poetic allegory of Spain conceived by its designer, the architect Josep María Jujol, and through the duality of two antagonistic elements, water, and fire, united by a single whole (Casajoana 2019). In keeping with the large monumental ensembles so typical of the period, the enormous triangular base has at its apex three large bronze sculptural groups representing the fruits and gifts of the waters (Abundance, Public Health and Navigation); both of the seas are represented in the three pools that are distributed on the sides of the triangle (the Mediterranean, the Cantabrian and the Atlantic Ocean) into which the waters of the main rivers of Spain flow (these organised into three allegorical groups according to sea in where they end). The three sculptural groups dedicated to the rivers are housed in three niches in the enormous architectural pedestal, which in turn is delimited at its three corners by giant columns dedicated to religion, heroism, and the arts. The monument is crowned by an enormous cauldron where a flame would burn forever, symbolising "the constant sacrifice of Spain for civilisation" (Grandas 1987, 235–7). Miguel Blay's large sculptural group, dedicated to the rivers of the Atlantic and correctly oriented towards the south, is marked by the presence of two large male figures: the Tagus (represented as a bearded adult man sitting on the edge of the rocky base, holding a large singing bowl), and the Guadalquivir (standing, beardless, although mature, and supporting a large shell where the water of one of its tributaries, the Guadalimar, pours out, personified in the form of a child carrying a singing bowl). Between them, appears a kneeling adolescent pouring its water into the Atlantic through a tile, the Guadiana River. The Ibero-American exhibition in Seville was also no stranger to this and had another monumental fountain, more discreet and dismembered after the exhibition. In this one, through fluvial allegories, there is another idea in vogue at the time: "La Hispanidad" (the feeling of union among all countries historically related to Spain). The so-called Fountain of Hispanity, was designed by the architect José Granados de la Vega, and contained the allegories of Iberia presiding over the fountain (by Francisco Marcos Díaz Pintado), and on either side two allegories: the river Magdalena on the left, representing the American rivers (made by José Lafita), and the river Guadalquivir for the Spanish rivers (by Agustín Sánchez Cid). The aforementioned allegory of the former Betis (which like the rest of the sculptural pieces in the group went to the María Luisa park after it was dismantled) shows the essential characteristics of the iconography of the river in the city of Seville: depicts it by lying on a bull's head and bearded, also recovered the horn of plenty, with the traditional urn through which the water poured (replaced by a spout that was located under the pedestal where it was erected). The work has evident classical influences such as the ripples of the waters where it stands, or the powerful anatomy, which evokes Michelangelo's models.

The arrival of the avant-garde meant that traditional interpretations coexisted with absolutely novel and heterodox ones, such as the Guadalquivir woman by the illustrator and designer Romain de Tirtoff (also known as Erté). With his Art Dèco style, he shows us the personification of the city of Seville as a graceful woman dressed in a long tailed gown, which turns into the blue waters of the river. The figure has a comb, from which hangs a mantilla that has become a net that is directed towards an Atlantic Ocean full of fish (these are being collected in an enormous basket, carried by a saleswoman on her head). Nevertheless, the traditional iconography was sufficiently established in its master lines. This is why in the Seville of the eighties of the twentieth century, the Tarife-born Guillermo Pérez Villalta returned to it in the enormous oil painting on the awning of the patio of the Seville Chamber of Commerce, Industry and Navigation building, which represents an enormous allegory of the institution. In it, the celestial vault serves as a frame for the compass rose, while it is surrounded by a frieze of mythological figures (this seems to evoke the apotheosis of Hercules painted by Francisco Pacheco in 1604 for the ceilings of the Casa Pilatos). In the upper left-hand corner, the Guadalquivir appears in the appearance of the traditional bearded man, paired with Neptune, picking up a small galleon that the god of the seas gives him (trade between America and the port of Seville). The iconography of the river is completed by the tower of the Giralda and some lake plants.

Finally, contemporary numismatics created singular examples. The Madrid painter and sculptor Fernando Somoza did an extensive series dedicated to the rivers of Spain. Demonstrating a personal style based on a firm drawing and synthesised forms, the sculptor finds the images of his rivers in the most varied sources (traditional mythology, popular scenes, literature, and the history of the regions they bathe). He uses allegory in the personifications of the Guadalquivir and the Guadiana, presenting them as old men with long beards who metamorphose into water, although the former is presented in a more naturalistic and traditional pose, while the latter is interpreted in a totally symmetrical way. With the Huelva rivers Tinto and Odiel, he abandons the allegory by placing a scene alluding to the Columbian epic, with four figures, perhaps including Columbus himself and the Pinzón brothers, and behind them the masts and sails of the three ships.

As a result of the contemporary urban growth of cities in Extremadura on the banks of the Guadiana, we find new allegorical examples with this river in more recent times. The representations of the Guadiana made in Andalusia are meagre, if not non-existent. In the Fountain of the Guadiana that Enrique Pérez Comendador gave to the town of Don Benito in 1967 (Bazán 2010, 220), the sculptor starts from the aforementioned Nilotic allegory of the Vatican museums, oscillating in representing the male image of the river between hairless and bearded, maintaining the traditional reclining position, as can be recognised in the preparatory sketches. He opted for the second version: the river lying on a fragment of column shaft, some coins (symbol of the riches emanating from the river) at the base and the fish in his hands. It has its eyes closed, following the verses of Celestino Vega Matos which are inscribed on its base: "Long laziness stretched out, river that dies and is reborn, dream of skies and waters. Guadiana, slow Guadiana". The other piece (by Juan de Avalos) is the sculptural group of the city on the river inaugurated in 1999 in Badajoz: a monumental fountain that presents a female figure as an allegory of the city that pours its powerful fountain as a male figure (the river Guadiana), beardless but mature, is seated on the ground, and which in turn carries a container from where it pours the liquid element into the main basin of the monument.

The humanisation of Andalusian rivers is a phenomenon that has extended into the twenty-first century. One of the most significant is the Fuente de las

Granadas (Fountain of the Pomegranates), created in 2007 by Ramiro Megías and located in Granada's Paseo del Salón. It is a monument cast in bronze by the Moliné Segovia Foundry, it measures eleven metres high and 15 metres in diameter. Its composition is organised on three levels. On the first, thanks to four large sculptures, the Sierra Nevada can be seen, wearing a snow flower in her hair, and La Vega, as a woman with an ear of wheat, as well as the Darro River, with a plate, evoking the pans for washing gold, and the Genil river. On the second level there are some pomegranates, while on the third a half-naked woman with a pomegranate in her hand and positioned looking in the direction of the fountain of battles. A particular feature of the iconographic representation is that the figures are standing, not reclined.

Finally, it is important to mention two other sculptures that represent allegories of rivers. The first is a sculpture made by Carlos Delgado in 2019, which represents an allegory of the Tinto River. It is a work in which there is no traditional human figure. It was donated by the sculptor to Riotinto. The second is a ceramic panel representing an allegory of Coria del Río and the Guadalquivir River. "The work is an allegory of Coria del Río and the Guadalquivir, images that were placed in an ephemeral triumphal arch that was erected in 1570 on the occasion of Philip II's visit to Seville. In the middle, there is a Phoenician coin with the name of Caura and a shad, an ancient fish that has disappeared from the river" (Diario de Sevilla, 17 November 2021).

References

Álvarez Martínez, J. M. 2020. "El culto a las aguas en Augusta Emérita." BRAEX (*Boletín de la Real Academia de Extremadura de las Letras y las Artes*) 28: 9–34.

Anónimo. 1924. *Inventario de las obras de arte* [...] *del Museo del Excelentísimo Sr. D. Enrique de Aguilera y Gamboa, XVII Marqués de Cerralbo*, signed in 1924 by Juan Cabré.

Anónimo. 2021. "Una alegoría sobre Coria del Rio y el Guadalquivir sustituye a un mural franquista." *Diario de Sevilla*, 17 November 2021 [online]. Accessed 16 March 2023. https://www.diariodesevilla.es/provincia/alegoria-Coria-Rio-Guadalquivir-sustituye-mural-franquista-memoria-historica_0_1630037420.html.

Bazán de Huerta, M. 2010. "La escultura monumental de Enrique Pérez Comendador." *Norba: Revista de arte* 30: 197–226.

Brewster, H. 1997. *The River Gods of Greece: Myths and Mountain Waters in the Hellenic World.* Londres: I. B. Tauris.

Canto, A. M., A. Bejarano, and F. Palma. 1997. "El mausoleo del dintel de los ríos de Mérida, Revve Anabaraecus y el culto de la confluencia." *Madrider Mitteilungen* 38: 247–94.

Casajoana Salvi, P. 2019. *Jujol versus Jujol: la fontana de la Plaça d'Espanya (1928–29)*. Badalona: Pere Casajoana Arquitectes.

Ciuccarelli, M. R. 2006. "Acheloo ctonio dalla Magna Grecia all'Etruria?" *Mediterranea* 3, 121–41.

Costa Solé, A., L. Palahí Grimal, and D. Vivó I Codina (eds.). 2011. *Aquae Sacrae. Agua y sacralidad en la Antigüedad*. Gerona: Universidad de Gerona.

De Mal Lara, J. [1570] 1992. *Recibimiento que hizo la muy noble y muy leal ciudad de Sevilla a la C.R.M. del Rey don Felipe N.S.* Seville: Universidad de Sevilla.

Fernández-Paradas, A. R. and R. Sánchez-Guzmán. 2021. "Uniendo caminos, contando historias: el puente en España en la Época Moderna, valores simbólicos e iconográficos." *Revista Histoira* 26 (1): 203–39. https://doi.org/10.5216/hr.v26i1.68275.

Gais, R. M. 1978. "Some problems of River-God Iconography." *American Journal of Archaeology* 82 (3): 355–70. https://doi.org/10.2307/504463.

Gómez Mayordomo, A. 2014–2015. "Iconografía el dios-río Nilo y su presencia en la musivaria hispanorromana." *Anas* (27–28): 87–105.

Gómez Mayordomo, A. 2018. "Iconografía de los dioses-río: representaciones en el arte musivo romano." Unpublished Master's Thesis, Complutense University of Madrid.

Gómez Mayordomo, A. 2019. "Los ríos del Paraíso iconografía y valor sacro en el cristianismo." *Revista digital de iconografía* 21: 55–86.

Granados Chiguer. I. 2019. "Introducción al estudio del mosaico mitológico romano en Andalucía". Final Degree Project, University of Cadiz.

Grandas, M. C. 1987. "Los proyectos urbanísticos para la plaza de España." *D'art* 13: 225–40.

Herrero, F. 1995. *Recorrido mitológico. Por las Fuentes de los Jardines de La Granja.* Madrid: Editorial Biblioteca Ícaro.

Hesíodo. 1978. *Teogonía*, introducción, traducción y notas de Aurelio Pérez Jiménez y Alfonso Martínez Díez. Madrid: Gredos.

Jentel, M. O. 1987. "La représentation du dieu Nil. Sur les peintures et les mosaiques et leur contexte architectural." *Echos du Monde Classique. Classical Views* XXXI (6): 209–16.

León Alonso P. 1995. *Esculturas de Itálica.* Seville: Empresa Pública de Gestión de Programas Culturales.

Loza Azuaga. M. L. 1993a. "Estatuas - fuentes romanas de Colonia Patricia Corduba." *Anales de arqueología cordobesa* 4: 141–58. https://doi.org/10.21071/aac.v0i.11381.

Loza Azuaga. M. L. 1993b. "La escultura de fuentes en Hispania: ejemplos de la Baetica." In *Actas de la I Reunión sobre Escultura Romana en Hispania*, edited by T. Nogales Basarrate, 97–112. Madrid: Ministerio de Cultura.

Martín Sedeño, S. 1825. *Compendio histórico, topográfico y mitológico de los jardines y fuentes del real sitio de San Ildefonso.* Madrid: Imprenta de Don Antonio Martínez.

Montero Herrero, S. 2012. *El emperador y los ríos. Religión, ingeniería y política en el Imperio Romano.* Madrid: Universidad Nacional de Educación a Distancia.

Moreno Arana, J. A. 2014. "Apologías políticas, erudición y enseñanza del latín: el emblema de Diego de Castro." *Revista de Historia de Jerez* (16–17): 151–68.

Morillas Alcázar, J. M. 1991. "El puerto de Sevilla en pinturas y grabados del siglo XVIII". In *Andalucía, América y el mar: Actas de las IX Jornadas de Andalucía y América*, edited by B. Torres Ramírez, 305–322. Sevilla: Diputación de Huelva.

Nogales Basarrate, T. 2002. "Aquae emeritenses: monumentos e imágenes del mundo acuático en Augusta Emérita." *Empúries* 53: 102–4.

Ostrowski, J. A. 1991. *Personifications of rivers in Greek and Roman art*. Cracovia: Uniw. Jagielloński, Państwowe Wydawnictwo Naukowe.

Pérez Sánchez, A. 2012. *Inventario de las pinturas de la Real Academia de Bellas Artes de San Fernando*. Alicante: Biblioteca Virtual Miguel de Cervantes.

Ponz, A. 1787. *Viage de España: en que se da noticia de las cosas más apreciables, y dignas de saberse, que hay en ella*. Madrid: Viuda de Ibarra, Hijos y Compañía.

Portús, J. 2016. *Metapintura. Un viaje a la idea del Arte*. Madrid: Museo Nacional del Prado.

Quiles, F. 2007. "José María Arango (1790–1835), Pintor. Una voz en el desierto." *ACADEMIA. Boletín de la Real Academia de Bellas Artes de San Fernando* (104–105): 107–32.

Rodríguez López, I. 2002. "Poseidón y el Tiazos marino en el arte mediterráneo: (desde sus orígenes hasta el S. XVI)". PhD diss., Complutense University of Madrid.

Salvá Herán A. 1997. *Colecciones artísticas del Congreso de los Diputados*. Madrid: Congreso de los Diputados, Fundación Argentaria.

San Nicolás Pedraz, M. P. 1997. "Representaciones alegóricas de fuentes y ríos en los mosaicos romanos de Hispania." In *Termalismo antiguo*, edited by UNED – Universidad Nacional de Educación a Distancia, 467–79. Madrid: UNED.

San Nicolás Pedraz, M. P. 2004–2005. "Seres mitológicos y figuras alegóricas en los mosaicos romanos de *Hispania* en relación con el agua". *Espacio, tiempo y forma. Serie II, Historia antigua* 17–18: 301–34. https://doi.org/10.5944/etfii.17-18.2004.4429.

Santiago Páez, E., and M. Cera Brea (eds.). 2016. *Ceán Bermúdez. Historiador del arte y coleccionista ilustrado*. Madrid: Biblioteca Nacional de España.

Swetnam-Burland, M. 2009. "The Vatican Nile." *American Journal of Archaeology* 113 (3): 439–57. https://doi.org/10.3764/aja.113.3.439.

Torres Carro, M. 1990. "Iconografía marina." In *Mosaicos romanos: estudio sobre iconografía: Actas del homenaje "in memoriam" de Alberto BalilIllana que tuvo lugar en el Museo de Guadalajara los días 27 y 28 de abril de 1990*, edited by D. Fernández Galiano, 107–35. Guadalajara: Asociación Española del Mosaico.

Zúñiga, L. [1671] 1747. *Anales eclesiásticos y seculares de la M. N. y M. L. ciudad de Sevilla*. Sevilla.

Index

Abastecimiento de Aguas Potables de Jerez de la Frontera S. A. 73
Abengoa 10
A Coruña 70, 96
Actuaciones para la Gestión y la Utilización del Agua (Actions for the Management and Use of Water) 23
Adam, John Mackay 141
Administração Geral dos Serviços Hidráulicos e Eléctricos (Department of Hydraulic and Electric Works) 46, 47
Aguadores 81, 81*n*1, 82
Aguas de Barcelona 10, 11, 28
Aguas de Cádiz S.A. 199, 201
Aguas de Gévora 53
Aguas de Guadix 10
Aguas de Herrera 10
Aguas de Huelva 54, 199, 201
Aguas de la Cuenca del Guadalquivir 5
Aguas de la Cuenca del Segura 5
Aguas de la Cuenca del Sur 5
Aguas de la Trinidad 150
Aguas del Huesna 10
Aguas de Linares 53
Aguas Potables de Albacete 72
Aguas Potables de Cádiz 53, 74
Aguas Potables de Córdoba 53, 73
Aguasvira 10
Albacete 71
Alcalá de Guadaira 123, 126, 130, 132, 138, 166
Algeciras 54, 73, 79, 123
Alicante 41, 70, 96, 122, 123
Aljarafe 162–165, 165*n*, 166, 168–170, 170*n*2, 174, 175, 177–179, 179*n*4, 180, 182
ALJARAFESA 162, 163, 163*n*, 164–167, 169, 171, 182
Allegory 225, 226, 232, 234–236, 238–242
Almensilla 164*n*, 165, 167, 170*n*2
Almeria 4, 9, 10, 73, 115*n*, 199, 199*n*, 201–204, 206, 219–221
Alta Comissão para Melhoramentos Públicos (High Commission for Public Improvements) 44
Andalusia 1, 2, 4, 6–12, 37, 53, 60, 68, 72, 78, 79, 103–105, 109, 118, 123, 124, 135, 146, 162, 166, 175, 187, 188, 193, 197, 198, 198*n*1, 199–201, 204, 206, 225, 228, 231, 232, 241
Andalusia Water Company Limited, S. A. (Andalusia Water C. Ltd.) 54, 73
Antequera 103–109, 111, 114–116, 118, 154*n*
Añarbe River 71
Aquagest 10, 28
Aqueduct 81, 82, 91, 95, 98, 99, 109, 112, 113, 116, 123, 132, 150, 151, 154, 156, 159, 168n
Aqualia 10, 28, 199, 201
Archaeological Museum of Cordoba 228
Archaeological Museum of Seville 228
Argentina 216
Astosam 10
Australia 216, 217, 220, 221
Austria 216
Autonomous Region 5–7, 9, 20, 23, 25, 193, 197, 198*n*
Aznalcázar 163–166, 176, 176*n*5, 176*n*6, 177, 178*n*4, 179*n*4, 180
Aznalcóllar 163, 163*n*, 166, 167

Baltá, Narciso 82
Banca Marsans S.A. 155
Banco de Crédito Local 117
Barcelona 10, 11, 28, 50, 70, 96, 100, 108, 122, 155, 175, 176*n*1, 178, 239
Barnett & Gale 90
Bell, Valentín G. 90, 91
Benacazón 164–167, 170, 171, 175–179, 175*n*2, 176*n*1, 177*n*3, 178*n*1, 178*n*2, 179*n*1, 179*n*2, 179*n*4
Bergamín, Francisco 152
Bibliometric analysis 210–212, 214, 216, 221
Bilbao 71
Binomial tariff 194, 195, 205
Birmingham 66
Blue gold 15
Bollulos de la Mitación 164, 166, 167
Bonuses 203–205
Bores Romero 144, 154–159
Brazil 216, 217, 220, 221
Brioso Raggio, Tomás 115, 115*n*, 116, 117
British foreign investment 122

Cadiz 4, 8, 9, 53, 74, 78–84, 84*n*, 85*n*2, 86, 87, 89, 90–98, 90*n*1, 98*n*, 99, 100, 123, 199, 199*n*, 201–204, 207
Caletilla 99
Canada 217
Canal de Isabel II 28, 72, 151,
Caños de Carmona 123, 127, 132
Carreira Gallardo, José 107
Carrión de los Céspedes 163–167, 169*n*2, 172, 172*n*3, 172*n*4, 172*n*5, 173, 176, 178, 179*n*4
Castilleja de Guzmán 164–167
Castilleja de la Cuesta 164, 165, 167, 182
Castilleja del Campo 163, 166, 167
Castilleja de Talhara 164
Central Commission of Local Sanitation 117
Cesspools 62, 150*n*
Chile 3, 217, 219, 220
China 210, 216, 217, 220, 221,
City of Las Palmas Water and Company Limited 71
Civil War 51, 93, 115*n*, 129, 132
Companies 1, 4, 5, 7, 9–12, 17, 19, 24, 26–29, 37, 38, 52–55, 61–66, 70–73, 78–80, 82, 95, 97, 99, 107, 109, 111, 121–124, 126, 128, 133, 135, 139, 146, 148, 149, 152, 153, 158, 190, 199, 200
Compañía Anónima de Saneamiento y Urbanización de Sevilla 128
Compañía de los ferrocarriles de Sevilla-Jerez-Cádiz 91, 95, 99
Concejo 173, 174
Concerted management 7, 190
Concession 6–8, 10, 11, 24, 27, 37, 39, 66, 70–74, 84–86, 85*n*1, 88, 89, 97, 100, 107, 108, 111, 123, 124, 124*n*, 127, 128, 131, 133, 135, 136, 138, 146, 148, 149, 151–153, 190, 192
Confederación Hidrográfica del Guadalquivir 133
Congresso Municipalista Alentejano (Alentejo Municipalities Congress) 43
Contamination 62, 123, 145, 146, 150, 157, 169, 176–179, 177*n*3
Cora sevillana 163
Conselho de Melhoramentos Sanitários (Sanitary Improvement Council) 40
Conselho Superior de Higiene e Obras Públicas (Council for Hygiene and Public Works) 47
Contribución Territorial 108
Consumption 5, 11, 15, 25, 26, 30, 39, 42, 52, 72, 74, 80, 83, 93, 109, 110, 113, 115, 116, 126, 129, 130, 134, 144–147, 169, 175, 178, 187–189, 192, 193–197, 199, 201–207, 211, 222
Consumption fee 194
Cooperativa Eléctrica Antequerana 107
Cordoba 4, 8, 9, 53, 73, 108, 114, 163, 199, 199*n*, 201–204, 207, 226, 228, 234, 237, 238, 239
Coventry 66
CRIDESA 10

Decentralisation 5, 20, 41, 43, 45, 55, 67
Deffarque 114
Dehesa de la Vega 98
De la Escosura, Luis 83, 88, 94
Del Cacho, Matías 74, 83, 84, 84*n*, 85, 88–91, 88*n*, 90*n*1
Demographic density 162, 164, 165,
Demographic growth 48, 106, 110, 121, 149, 154, 162, 165–167
Denmark 216, 217, 220, 221
De Prado, José F. 87
Deregulation 3, 3*n*, 26
De Rojas y Arrese-Rojas, José 116, 117
Desalination 16, 21, 23, 26, 145, 189, 217, 218, 221, 222
De Torres, Francisco 113
Dickinson, William 141
Dictatorship 19, 29, 43, 44, 50, 53, 109, 115, 115*n*, 116, 117, 136*n*2, 177*n*1
Direcção Geral dos Edifícios e Monumentos Nacionais (Department of Buildings and National Monuments) 46
Dirección General de Sanidad (Health Directorate General) 40
Direct management 1, 7–11, 38, 50, 190
Distribution network(s) 8, 9, 52, 53, 73, 130, 132, 133, 137, 1369, 140, 144, 149–151, 155, 157–159
Domestic water supply 53, 69, 187, 199, 203
Dragados y Construcciones 10
Drinking water 1, 2, 4, 8, 10, 28, 49, 51, 60–66, 68–73, 75, 78–80, 82, 92, 94, 95, 97, 98, 100, 104, 111, 112, 115, 130, 146, 147, 149, 150, 155, 174–176, 178, 189, 199, 200
Duero 16, 237
Dupuy de Lôme, Enrique 155

Easton Amot & Hone 83
Ebro 16, 23, 225, 237
Ecological crisis 121
Economies of scale 1, 52, 65, 122, 126, 139
Efficiency 1, 3, 19, 20, 22, 26, 45, 56, 130, 131, 137, 139, 187, 188, 190, 191, 195*n*, 196, 197, 205, 222
Electricity 2, 11, 19, 41, 42, 64, 97, 107, 109, 115*n*, 121, 133, 212, 214, 215, 220, 245
Electric lighting 42, 73, 107
El Puerto de Santa Maria 81–83, 86, 87, 90–92, 95, 98
EMACSA (Empresa Municipal de Aguas de Córdoba, S.A.) 199, 201
EMASA (Empresa Municipal de Aguas de Málaga) 200, 201, 204
EMASAGRA (Empresa Municipal de Aguas de Granada) 10, 199, 201
EMASESA (Empresa de Abastecimiento y Saneamiento de Aguas de Sevilla S.A.) 200, 201
Empresa de Aguas Potables S. A. 73
Empresa de la traída de aguas de los manantiales de La Piedad a Cádiz 84
Empresa General de Construcciones 155
Empresa Pública Gestión de Infraestructuras, S.A. (GIASA) 7
Empresa Pública para el Desarrollo Agrario y Pesquero de Andalucía, S.A. (DAP) 7
Endesa 10
English investment 78
Environmental management 210, 218, 222
Environmental taxes 210–215, 221, 222
Estado Novo 38, 44
Espartinas 164–167, 176–178, 177*n*1, 177*n*3, 178*n*3
Estatuto do Trabalho Nacional (National Work Statute) 48
European Union (EU) 5, 16, 17, 27, 29, 145, 188, 198, 221

Federação de Municípios Alentejanos (Federation of Alentejo Municipalities) 43
Felipe Herrero, Bernardo Bouderé y Sobrinos 107
Ferrovial 10
Financial entities 55
Financialisation 26
Fomento de Construcciones y Contratas (FCC) 10, 28, 29, 199, 201
Foreign British investment 121
Foreign capital 12, 73, 97, 121, 122, 131, 135, 158
Foreign Office 133–137, 136*n*1
Forrect, Guillermo 90
Fountain(s) 45, 52, 53, 60, 61, 63, 71, 95, 99, 110, 111, 150, 169, 171, 172, 174, 176*n*4, 176*n*6, 178–180, 228, 230, 234, 236, 237, 239, 240, 241, 242
France 17, 28*n*2, 67, 75, 97, 104, 220
Franco 19, 29, 135, 139, 177*n*1
Francisco Menoyo y Compañía 82, 83
Free standing company 121, 124, 139

Garay, Joaquín R 126
García Carrera, José 117
Gas 42, 63, 64, 66, 73, 97, 106*n*, 107, 109, 121, 212, 214, 215, 221
Gelves 163–165, 169, 176, 176*n*2
Germany 104, 131, 216–218, 221
Gerena 162, 163, 166, 167
Gestión y Técnicas del Agua (Gestagua) 10, 29
Gines 164, 165, 167
Glyn, Mills, Currie and Co. 125
Granada 4, 8–10, 112*n*1, 112*n*2, 199, 199*n*, 200–204, 207, 226, 233, 234, 242
Great Britain 3, 61, 63, 67, 68, 75, 123, 131
Great War 43, 139
Greece 217, 227
Gros, Fréderic 151, 158
Guadalete 82, 83, 226, 234, 235
Guadalhorce 115*n*, 156, 226
Guadalquivir 5, 123, 124, 127, 130–133, 131*n*, 162, 163, 169, 226, 231, 232, 234–242
Guadiamar 162, 169
Guerrero Muñoz, Francisco 106

Halifax 66
Hamilton, John Montague 141
Hastings 66
Hawes, William 90
h-Index 213, 219, 220
Hidralia, Gestión Integral de Aguas de Andalucía S.A. 199, 200
Hidrocartera 10
Hidroguadiana 4
Hidroelétrica del Chorro 107

Higgin Winfield, Georges 123, 124, 124*n*
Hospital de San Juan de Dios 108
Huddersfield 66
Huelva 4, 54, 73, 115*n*, 123, 133, 163, 199–204, 199*n*, 207, 241
Huévar 164–166, 179*n*4
Hull 66
Hydrographic Confederations 20, 23
Hygienic conditions 41, 60, 75, 116, 181

Iberdrola 10
Iberian Peninsula 16, 17, 228, 236, 237
Iconography 225, 227–233, 236, 237, 239–241
Imperial Canal 71
Increasing blocks 201, 205
Indebtedness 55, 130,
Indias 216
Indirect management 3, 7, 9, 190, 199
Industrial Revolution 61, 63
Infrastructures 4, 6–9, 15, 16, 18, 25, 27, 35–39, 41–43, 47–56, 68, 103, 104, 109, 115, 115*n*, 118, 121, 123, 149, 152, 188, 191–193
Inima 10
Inland waters 22
Insalubrity 62, 67
Instituto Estudios de Administración Locales (Institute for Local Administration Studies) 51
Integral water cycle 187–189
Integrated System for National Hydraulic Balance (SIEHNA) 20, 21
Interested management 7, 190
Instrucción General de Salud (General Health Instruction) 40
Investment(s) 4, 6, 11, 25, 26, 28, 40–42, 46, 47, 49, 50, 52, 54, 74, 78, 85, 95, 97, 100, 109, 112, 117, 121–123, 130, 137, 158, 191, 211, 221
Ireland 216, 217, 220, 221
Isla Mayor 10, 163, 163*n*, 166–168
Israel 217, 220

Jackson, Frank S. 141
Jaen 4, 8–10, 73, 103*n*, 199–204, 199*n*, 206, 219–221, 232, 233
James Easton and Shaw 124
Jerez de la Frontera 73, 79, 81*n*2, 83, 96, 100, 235
JIF (Journal Impact Factor) 213, 217, 218, 222
Joint venture(s) 27, 28
José Abad y Compañía 82
José Bellido y Compañía 107
Junta Autónoma de Estradas (Autonomous Highways Board) 45
Junta de Propios 112

Kinnear, Benjamin G. 90

La Alameda 53
La Algaba 163, 166, 167, 182
La Constancia 54
La Cruz 85
La Estrella 54
La Florida 156
La Fraternidad 53
La Granja 234, 236
Lambert, Cowley 141
La Piedad 80, 83–87, 91, 92
Larios, José 82
Las Palmas 122, 123
Las Palmas de Gran Canaria 71
Leasing 7, 93, 94, 190
Lei de Reconstituição Económica (Law of Economic Reconstitution) 49
Liga de Propietarios de Fincas Rústicas (Association of Rural Property Owners) 122
Lisbon 37, 38, 45, 48, 49, 51, 55, 148*n*2
Liverpool 66
Llatas Duque, Francisco 181, 182*n*
London 65–67, 73, 78, 83, 89, 90, 99, 100, 123–125, 132, 133, 135, 136
López Gamarra y Nogueira, Antonio 114

Mairena del Aljarafe 164*n*, 165–167, 167*n*, 182
Malaga 4, 8, 104, 108, 114, 115, 115*n*, 144, 149, 151–155, 151*n*2, 153*n*2, 158, 201–204, 207, 226
Madoz 108
Madrid 43, 51, 72, 83, 108, 115*n*, 133, 134, 136, 151, 155, 175, 232*n*, 235, 235*n*1, 241
Magdalena 111–114, 112*n*1, 112*n*2, 116, 118, 240
Management 1–3, 5, 7–12, 15–17, 19–30, 35, 37, 38, 43, 44, 50, 54, 55, 61, 63, 68–73, 78–80, 99, 107, 115, 118, 121–126, 139, 144–149, 151, 153, 158, 159, 162, 163*n*, 168,

170, 176*n*6, 187–192, 195, 197–201, 205, 210, 211, 218, 222
Manchester 64, 66
Marbella 10
Martínez Rubio, Félix 114
Mexico 172*n*3, 217, 220
Mixed companies 7, 9, 10, 190
Modern Age 106, 109, 111
Modern Drinking Water System 79, 146, 147
Molina Lario 150
Monomial tariffs 196
Monopoly 1–3, 27, 37, 65, 107, 111, 189–191
Montañés y Álvarez, Diego Fernando 74, 97, 97*n*,
Moreno Burgos, José 112, 113, 113*n*
Moreno Fernández de Rodas, Carlos 116, 117
Morer Abril, José 151
Municipal by-laws 113, 199–205, 199*n*
Municipal Health Regulations 176
Municipalisation 12, 35, 37, 38, 41, 55, 56, 70, 72, 73, 79, 121, 123, 131–133, 135, 140, 144, 149, 152–154, 158
Municipalities 4, 7–11, 24, 25, 27–29, 36, 39–44, 50, 52, 54–56, 60, 64, 66, 68–70, 72, 81*n*, 109–111, 115, 148, 163–168, 164*n*, 177, 179, 179*n*4, 189, 190, 197
Municipal industrialism 63, 64
Municipal Law of 1877 110
Municipal Sanitation Regulation 69
Municipal Trading 61, 63, 64, 70
Museum of Fine Arts in Granada 233

Nationalisation 2, 3
National Hydrological Plan 20, 22, 23
National Plan for Adaptation to Climate Change (PNACC) 19
National Water Supply and Sanitation Plan 12
Netherlands 216–218, 221
New Water Culture 16, 19
Nineteenth century 1, 4, 18, 36–39, 45, 54, 55, 60, 61, 63, 64, 66, 68, 74, 75, 78–82, 96, 99, 104, 105, 110, 111, 113, 121–123, 129–131, 139, 146, 149, 150, 150*n*, 165, 175, 179*n*4, 237
Norway 216, 217, 221

OHL 10
Old Regime 108
Olletas 156, 157
Ondagua 10
Oviedo 71
Oxford 66

Palma de Mallorca 71
Palomares del Río 163, 166, 167
Paris 27, 67, 151, 152,
Parejo Obregón y Rojas, Luis María 107
Partington, Guillermo 83
Personification 226–231, 233, 234, 237, 240, 241
Pilas 163–166, 175, 179*n*4, 182
Pinilla Aranda, Esteban 117
Pipes 11, 52, 64, 69, 73, 84, 86, 91, 94, 95, 97, 112–114, 116, 117, 130, 133, 145, 157, 162
Plymouth 66
Porto 37, 38, 40, 45, 48, 49, 51, 148*n*2
Portugal 23, 35, 37–40, 43, 44, 49–51, 54–56, 148*n*2, 217, 219
Primo de Rivera, Miguel 109, 115, 115*n*, 116, 136*n*2
Private companies 1, 4, 9, 11, 12, 24, 26–28, 54, 55, 61–64, 66, 70, 73, 78–80, 109, 111, 146, 148, 149, 153, 190, 200
Private management 2, 9, 27–29, 35, 37, 144, 149, 158, 190
Privatisation 2, 3, 8, 12, 15, 16, 26–28, 147, 149, 152, 190
Productivity 106, 211–213
Provision of public untaxed assets 200
Public company 7, 9, 28
Public health 6, 36, 38, 40, 41, 54, 60, 61*n*, 65–68, 75, 121, 181, 239
Public intervention 36, 121, 189
Public service(s) 1–4, 7, 8, 10, 11, 46, 60, 61*n*, 68, 79, 84, 95, 97, 103, 104, 106, 108, 109, 111, 114–116, 118, 122–124, 128, 131, 135, 139, 146, 147, 149, 150, 190
Public water supply 60, 78, 144, 163
Puerto Real 92, 95

Quartile(s) 164, 213, 217, 218, 222
Quay 87, 90, 95, 99, 100

Railway(s) 49, 82, 91, 92, 95, 97, 98, 108, 121, 123, 154*n*
Rainwater 80, 116, 117, 145, 170*n*2, 199
Ramos Gaitero, José 114

Reglamento para la Higiene y Salubridad de la Ciudad Alicante (Regulation of Hygiene and Salubrity of the City of Alicante) 41
Regulamento Geral de Saúde e Beneficência Pública (General Regulations for Health and Public Welfare) 41
Regulation(s) 3*n*, 5, 6, 8, 15, 16, 19–22, 24, 25, 29, 30, 37, 40–42, 42*n*, 44, 61, 66, 68, 69, 108, 111, 113, 121, 130, 176, 177, 181, 193, 198, 199, 200, 221
Renewable energies 212
River(s) 16, 17, 20, 23, 24, 52, 62, 71, 72, 82, 83, 109, 111–116, 115*n*, 118, 123, 124, 130–133, 136*n*2, 145, 150, 156, 162, 169, 180*n*1, 189, 225–237, 226*n*, 239–242
Robarts Lubbock & C° 90
Royal Council of Castile 112
Ruiz de la Herranz, Augusto 114
Rural improvements 45, 47
RWE Thames Water 10

Salamanca 71
Sales 10, 29, 85, 85*n*1, 88, 90, 93
Salteras 164, 166, 173, 174, 177, 180, 181
San Cristóbal 90, 98
San Fernando 10, 82, 95, 233, 234
Sanitary movement 60–63, 66–68, 75
Sanitary reform 62
Sanitary Report 62, 63, 65
Sanitation and Improvement Projects of Antequera 116
Sanlúcar la Mayor 163, 164, 166, 170*n*2, 176, 182,
Santiponce 165, 167, 168*n*
Santos Atienza 117
Sanz de Melgarejo, Diego 107
Service fee 194
Servicio Municipal de Aguas 139
Servicio Municipal de Aguas Potables de Córdoba 73
Sevillana de electricidad 138
Seville 8, 9, 96, 123, 134, 137, 162
Seville Waterworks (*Seville Water Works C° Ld*) 53, 73, 121, 123–125, 124*n*, 132, 139
Sewerage 7, 27, 36, 61, 62, 69, 73, 144, 155
Sixteenth century 150, 230–232
SJR (SCImago Journal Rank) 213, 217, 218, 222
Social interest 191
Sociedad de la Bahía de Algeciras 73
Sociedad General de Aguas de Barcelona 11
Southampton 66
Spain 2–5, 8–11, 15–21, 23–29, 28*n*2, 35, 37–39, 41, 43, 50, 51, 53, 54, 56, 60, 66–68, 70, 72, 78, 79, 90, 95, 99, 104, 105, 109, 111, 121–123, 126, 131, 133, 135, 136, 139, 144, 147, 148, 155, 162, 166, 176*n*6, 187, 188, 190, 192, 197, 198, 198*n*1, 210, 215–217, 219–221, 231, 236, 237, 239–241
Spanish Association of Water Supply and Sanitation (AEAS) 27
Spanish Geological and Mining Institute 170
Springs 8, 52, 53, 69, 71–73, 80, 82–85, 87, 90, 95, 98–100, 116, 124, 127, 130, 144, 145, 147, 150, 152, 155, 156, 158, 162, 169
Stagnation 74, 106, 129, 130
State Water Companies 24
Sustainability 3, 18, 22, 144–146, 187, 188, 191, 195*n*, 196–198, 205, 210, 212, 218, 221, 222
Switzerland 216–218, 220, 221
Symbolism 45, 225

Tank(s) 8, 45, 69, 80, 82–84, 86–88, 90, 95, 97–99, 109, 116, 151, 153, 158, 159, 176, 178
Tariff structure(s) 188, 192, 194, 195, 197, 199–202
Taiwan 216, 217, 221
Taxation 92
Tempul 81, 81*n*2, 83
The Cadiz Water Works Company Limited 78–80, 89, 90, 97, 98*n*, 99
Tomares 163–167
Torremolinos Waters 152, 154
Twentieth century 2, 4, 9, 16, 20, 21, 35, 37, 39, 43, 50, 51, 53–55, 69, 71, 72, 104–106, 118, 121, 126, 129, 131, 147–149, 153, 162, 165–167, 169, 173–175, 177, 178, 225, 226, 232, 239, 240
Twenty-first century 9, 15, 26, 167, 226, 231, 241

Umbrete 165, 167
Undersecretary of Industry 136
Unión de Municipios Españoles – UME (Union of Spanish Municipalities) 44

Unión Patriótica 109, 116, 136*n*2
United Kingdom (UK) 66, 67, 133, 217, 218
United Nations 15, 18
United Nations Development Programme 145
United Nations Millennium Development Goals 221
Urbanisation 105, 148, 165, 169
Urban modernisation 35
Urban services 103, 121, 122
Urban network(s) 38, 42, 52, 54, 55
Urumea 71
USA 210, 216–221

Valencia 53
Valencina de la Concepción 164, 169*n*2, 172, 172*n*1, 174, 174*n*8
Villamanrique de la Condesa 163, 163*n*, 164, 166
Villanueva del Ariscal 164, 165, 167
Vitoria 71

Warehouses 87, 99, 138
Wastewater 169, 170, 175, 189, 199, 210–222
Water Act 15, 16, 18–22, 66
Water Agreement 114
Water Culture 16, 18, 19
Water cycle 5, 20, 24, 25, 187–190, 192–194, 198–200
Water Framework Directive 5, 21, 187, 188, 191, 192, 195, 198, 221
Water Law 5, 18, 19, 98, 111, 198
Water management 5, 7, 15–17, 19–24, 28–30, 78, 122, 124, 145, 146, 168, 170, 187, 191, 198, 210, 211, 218, 222
Water policy 5, 6, 17, 21, 22, 192*n*, 198
Water real(es) 85, 85*n*1, 88, 93
Water resources 6, 7, 15, 16, 20, 79, 80, 99, 137, 139, 140, 145, 162, 168, 168*n*, 174, 187–189, 191, 192, 205, 210, 216, 221
Water tariffs 187
Welfare State 3
Wells 8, 52, 53, 80, 82, 91, 98, 99, 109, 123, 128, 130–132, 137, 147, 150, 150*n*, 151, 154, 162, 169, 171–179, 174*n*7, 189
White, Thomas 82, 83
Worcester 66
World War 3, 73, 126, 130, 132, 134, 139, 148
WoS database 210, 212–214
Wright, John Arthur 90

Zaragoza 71, 154*n*

www.ingramcontent.com/pod-product-compliance
Lightning Source LLC
LaVergne TN
LVHW010625300926
849162LV00001B/68